DELTA PRODUCTS

CASE STUDY

Marjorie Leeson
Delta College

SCIENCE RESEARCH ASSOCIATES, INC.
Chicago, Palo Alto, Toronto
Henley-on-Thames, Sydney, Paris

A Subsidiary of IBM

Acquisition Editor	Terry Baransy
Project Editor	Sara Boyd
Designer	Carol Harris
Compositor/Illustrator	Publications Services

SRA is a registered trade name and trademark of Science Research Associates, Inc.

Library of Congress Catalog Number: 80-29412

ISBN 0-574-21288-4

10 9 8 7 6 5 4 3 2 1

CONTENTS

Preface   v

1   **INTRODUCTION TO DELTA PRODUCTS**   1
   BACKGROUND   2
   DELTA PRODUCTS   2
   THE HISTORY OF DATA PROCESSING AT DELTA
     PRODUCTS   5
   DELTA'S HARDWARE   5
   DELTA'S OPERATING SYSTEM   7
   LANGUAGES   8
   CONVERSION TO AN ONLINE SYSTEM   9
   THE DATA PROCESSING DEPARTMENT   9
   MAJOR CONCERNS   11

2   **DELTA'S PAYROLL SYSTEM**   17
   THE PAYROLL DEPARTMENT   18
   THE HISTORY OF THE PAYROLL SYSTEM   19
   A NEED FOR CHANGE: NEW CITY, STATE, AND
     FEDERAL REGULATIONS   20
   A REQUEST FOR A SYSTEMS STUDY IS MADE   20

3   **PLANNING THE DETAILED INVESTIGATION**   27
   BACKGROUND   28
   PLANNING THE STUDY   28
   PROCEDURES RUN UNDER THE PRESENT SYSTEM   29
     Display Program, Maintenance Programs,
     Report Program
   EMPLOYEE CLASSIFICATIONS   32
   PAYROLL REGISTER PROGRAM   33
   STOCK DEDUCTION REPORT   34
   CREDIT UNION REPORT   34
   PRINTING THE PAYCHECKS   34
   UPDATING THE PAYROLL MASTER FILE   35
   MONTHLY PROCEDURES   35
     Payroll Distribution Report
   QUARTERLY PROCEDURES   36
     FICA Report
     Quarterly Change Program

ANNUAL REPORTS   37
   W-2 Forms
GENERAL COMMENTS ON THE PAYROLL SYSTEM   37
SUMMARY   38

4   **DETERMINING I/O REQUIREMENTS**   43
   ADDITIONAL PAYROLL INFORMATION   44
   GENERAL GUIDELINES FOR DEVELOPING TH
     PAYROLL SYSTEM   47
   PROGRAMS NEEDED TO IMPLEMENT THE NEW
     SYSTEM   50

5   **DESIGNING REPORTS**   63
   STOCK DEDUCTION REPORT   65
   FORM 941a   66
   W-2 FORMS   67

6   **DESIGNING SOURCE DOCUMENTS**   73

7   **FORMATTING A CRT SCREEN**   75

8   **INPUT, PROCESSING, AND OUTPUT CONTROLS**   85

9   **DEVELOPING A HIERARCHY CHART AND A
    STRUCTURED PROGRAM FLOWCHART**   87
   DEVELOPING A HIERARCHY CHART   89
   DEVELOPING A PROGRAM FLOWCHART   90
   CONCLUSIONS REGARDING HIERARCHY AND
     PROGRAM FLOWCHARTS   94

10   **DOCUMENTING THE SYSTEM**   97

11   **IMPLEMENTING THE SYSTEM**   105

12   **PROVIDING IN-SERVICE TRAINING**   109

Can an analyst design a payroll system to meet the specific needs of Delta Products if he or she knows nothing about the company or its present system? Can an analyst design a system without understanding the characteristics of the hardware and software available for implementing the system? What about union contracts and the federal, state, and local laws that might have an impact on the payroll system? Obviously, before an analyst can design a system, he or she must understand the capabilities of existing hardware and software, be aware of the company's needs, and become acquainted with union contracts and management policies.

The student programmer/analyst is given background information on Delta Products and its management, the computer system available for implementing the payroll system, and the present payroll system. The student also discovers that the recent passage of several new tax laws will require major changes in Delta's payroll system. Mr. Paul, the data processing manager, receives a request from Mary Smith, Delta Products controller, for a study to determine the feasibility of designing a new payroll system.

After the computer policy committee has reviewed the request, two analysts, Arnold and Walczak, are immediately assigned to the project. When the preliminary investigation report is received, the committee recommends that Arnold and Walczak begin work on a detailed investigation that will produce a general design for a new system.

At this point, the student, who will do many of the tasks normally completed by analysts, programmers, or operations personnel, is asked to identify the problems with the present system and make recommendations for the design of the new system.

For the new payroll system, the student is asked to complete the following tasks:

1. Determine the output requirements.

2. Determine the input requirements.

3. Determine where and how data should enter the system.

4. Determine the contents of the payroll master file, the payroll transaction file, and the output files used to produce reports.

5. Determine the procedures needed to provide the required output, maintain the payroll master file, and protect the integrity of data stored in the files.

6. Design source documents.

7. Design reports.

8. Format CRT screens.

9. Determine the input, processing, and output controls needed for each procedure.

10. Provide a written overview for the entire system and completely document a major procedure.

11. Determine which employees need an overview of the new system and which employees need specific training prior to the implementation of the new system.

12. Determine when and how the new system should be implemented.

Delta Products Case Study, although designed to accompany the author's *Systems Analysis and Design* textbook, contains enough detail and substance to be used as a supplemental learning tool with other textbooks. (The Solutions Manual explains how to coordinate the sections in the casebook with the appropriate background material in three other systems analysis textbooks.) The casebook was designed for use:

1. as supplemental learning material for a systems analysis and design course or a data processing applications course.
2. in an advanced programming course. Students can design and implement the system.
3. in independent study programs.

As you evaluate the casebook and the tasks the students are asked to perform, you will discover that the material in the casebook is:

Realistic   The material regarding Delta Products' present payroll system is realistic. Students must use current information on state and federal income tax laws and procedures (including FICA) in solving the tasks. Students complete print layout forms for actual reports submitted on 941a and W-2 forms.

Relevant   The hardware and software products available for implementing the payroll system at Delta Products represent the current state of the art in computer technology. Many companies are now changing from totally batch-oriented systems to ones that use online terminals and distributed data entry. Delta Products' new payroll system will include some online procedures.

Flexible   Depending upon the time available, the tasks assigned to students can be increased or decreased. Once the requirements of the payroll system are defined, future assignments can deal with any or all of the required procedures.

Stimulating   Just as an analyst must refer to and use data gathered during the investigation phases of a project, the students must refer to the background material provided in the first few assignments. They must dig out the facts needed to complete an assigned task.

Learner-oriented   Although students must sift out relevant data for themselves, we have removed as many obstacles as possible from the learning process. We give examples to illustrate how a particular assignment might be completed. Although students are expected to be creative in the design of the new system, we present guidelines to keep them on course.

Complete   All the material needed to perform the required tasks is included in the casebook. This includes both the forms, such as file and print layouts, and the specific information needed to design and implement the payroll system. The one exception was mentioned above: students are asked to obtain information from the library pertaining to their state's income tax laws and federal withholding tax laws.

Easy to use   Suggested solutions for all assigned tasks are given in the Solutions Manual. The manual also contains suggestions on how to use the material in an independent study environment or in an advanced programming class.

Just as the payroll system designed by the students must be evaluated on a continuing basis, so must this casebook. We believe that it will prove a valuable addition to your learning system.

INTRODUCTION TO DELTA PRODUCTS

1. To obtain background information on Delta Products' management, hardware, software, and data processing personnel.
2. To determine what type of problems faced Ben Paul, Delta Products' data processing manager, when he took over Delta's data processing department.

If you are unfamiliar with any of the following terms or phrases, look them up in the glossary of your text or in a data processing dictionary.

Assembler language	JECL
Batch jobs	Online processing
BPI	Operating system
Byte	Paging
CICS	Photoelectric card reading
Controller	POWER/VS
Control unit	Programmer/analyst
Database	Queue
Data exception	Real mode
Data module	Serial card reader
Direct access storage	Strategic planning
Distributed data entry	Structured design
DOS/VS	Structured programming
EDP	Supervisor
Emulated	SYSGEN
Formal organization	Throughput
IBG	Virtual mode
Informal organization	Virtual storage
Integrated monolithic memory	Virtual system
IPS	

PREPARATION   Before undertaking the tasks assigned, complete the following reading assignment: Chapter 1 (An Overview) and Chapter 2 (Management and the Data Processing Department). (*Note*: The above chapters are from Leeson's *Systems Analysis and Design* published by SRA. If you are not using that text, your instructor will tell you what to read.)

TASKS   1. Read the background information that follows.
2. After you have studied the material, answer the questions on pages 13–16.
3. Review the list of key words. If you are unfamiliar with any of the terms, look up their definitions.

BACKGROUND   Before a new system can be designed, the individuals responsible for the design must understand:

1. The organization, its management and its information needs.
2. The characteristics of the hardware to be used for implementing the design.
3. The characteristics of the control software (operating system) that is available or that will be obtained prior to implementing the system.
4. The reason a new system is needed.
5. The specific objectives of the new system.

Since you are to assume that you are either an analyst, a programmer, or a programmer/analyst for Delta Products, you should understand something about your company before you can determine what type of system should be designed.

DELTA PRODUCTS   Delta Products up to this point has been a rather conservative company that manufactures parts and subassemblies for the automotive industry. The company was initially organized by five engineers who once worked for General Motors.

Delta was organized and incorporated right after World War II, a time of extraordinary expansion in the automotive industry. Old companies were converting from wartime to peace-time production, and new businesses with sound financial backing grew at an unusually fast rate.

When the company was first organized, the five founding engineers had managerial positions within the company and also served on the board of directors. (This was understandable because the five initially owned, or controlled, the majority of the stock.) The five engineers were very competent technically but lacked experience in managing a rapidly expanding company.

Very soon after the company was incorporated (and the founders' lack of experience became evident), the engineers hired a management consultant firm to study their company. The consultants were asked to:

1. Determine the problems within the company that related to the administrative organization of the firm.

2. Determine what problems might develop if changes were not made.

3. Determine how marketing, production, and capital expenditure decisions were currently being made.

4. Ignore constraints such as lack of personnel or equipment.

5. Based upon the past experience of Delta Products and the projected growth of the Gross National Product, determine short-, intermediate-, and long-term projections for sales and the growth of their company. The projections were to cover the remainder of the forties and the entire decade of the fifties.

6. Based upon the projections, determine the organizational structure needed to manage the company effectively. If new positions were to be created, job descriptions were to be included in the recommendations.

7. Make positive, detailed recommendations regarding the changes needed to accomplish the projected growth.

8. Identify the areas that present the greatest possible danger, prioritize the danger areas, and then give recommendations for solving those problems with the highest priority.

The consultants did recommend an administrative reorganization and developed job descriptions for several key executives that would bring into the company the additional expertise (outside of the engineering and the product development area) it needed. It was also recommended that the individuals hired for the new top-management positions should be part of some type of profit-sharing plan. Since they would be responsible for much of the **strategic planning**, they should be given the same type of incentive to promote the growth and development of Delta Products that the five founding fathers possessed.

Their report also indicated that there were a number of problems that directly related to the failure to obtain the kind of relevant information needed in the decision-making process. For this reason, the consultants suggested that the company investigate the possibility of using punched card equipment to process some of their data. They also recommended that if Delta decided to use punched card equipment, a separate department should be created and placed under the jurisdiction of the vice-president of finance.

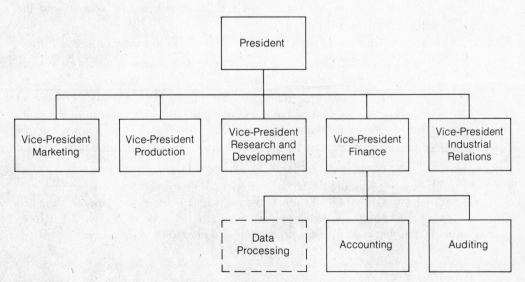

Suggested organization for Delta Products

Although the recommendation to develop a data processing department was given a high priority, it was not considered as critical as developing better internal controls and better defined procedures.

The consultants were in agreement that the two major areas that should be addressed immediately were the lack of:

1. well-defined top-level managers in some areas and
2. well-defined procedures.

The consultants felt that Delta Products had been lucky. At the time the company was formed it would have been virtually impossible to fail because the company had good financial backing and a ready-made market for their products.

Items such as standards manuals and procedural manuals did not exist. One could not cite the differences between an **informal** and **formal organization** as a problem since a formal organization did not exist. The company had expanded rapidly and had established few procedures or guidelines. Day-by-day problems were addressed and solved as they arose. Seldom were historical information and future projections used in the decision-making process.

The consultants also recommended that the president should be someone other than one of the company's founders. The five founders had been considering rotating the presidency. However, they did agree that the rationale for obtaining a president from outside the firm should be followed.

Many of the recommendations were implemented immediately, and a more formal organization began to emerge. In the late 1950s the company did a preliminary investigation into the methods used for processing data to determine if a detailed investigation was warranted. The preliminary investigation did show that the information needed for the decision-making process was not timely. Every manager in every functional area complained that the information was almost obsolete by the time it was received and had value chiefly as a historical reference. Marketing and sales were concerned about the number of customers who received the wrong products or who experienced delays in receiving orders. Upon reading the report, management was in favor of conducting a detailed investigation to determine if the company should obtain punched card equipment for processing routine, repetitive data.

The detailed investigation resulted in the following recommendations:

1. Lease punched card equipment.
2. Determine what systems can be cost-justified when redesigned and implemented on unit-record equipment.
3. Develop more automated general-ledger and cost-accounting procedures.
4. Establish a data processing department. The manager would be responsible to the vice-president of finance.
5. Hire a data processing manager who will be given the responsibility of establishing the data processing department, developing its long-range objectives, and for providing inservice training for top- and middle-management personnel who will be directly involved in the utilization of electronic data processing (**EDP**).

The analyst who did the study was responsible to research and development. He felt that ultimately Delta Products would want to obtain a computer. Punched card equipment, and the development of automated procedures, would help to bridge the gap between manual processing and electronic data processing. At the time the study was done (1958), it seemed as if computers were totally out of the question for companies like Delta Products. However, the analyst was firmly convinced that as technological changes occurred, computers would become available and could be cost-justified for companies as small as Delta Products.

The recommendations of the analyst were approved, and a data processing department was formed. In 1960 Delta's data was either processed manually or electromechanically by punched card equipment. Electromechanical equipment controlled by a wired control panel recorded, sorted, condensed, and calculated data punched into cards. A typical job required a great deal of card handling, and the equipment's calculating and printing capabilities were limited.

THE HISTORY OF DATA PROCESSING AT DELTA PRODUCTS

 The analyst who did the study in 1958 was right. In 1963 a second-generation computer that had a limited memory (8K), tape drives, a card read/punch unit, and a printer was acquired. It was programmed in an **assembler language** and did not have an **operating system**. Only batch jobs could be run, and **direct access storage** was not available since its use could not be cost-justified. There was still a great deal of card handling and some of the punched card equipment was retained in order to prepare input for the computerized system. The vendor provided training for the data processing manager and for the unit-record equipment operators who would operate the new computer.

 Since it was almost impossible to secure experienced programmers, two individuals within the firm were selected to receive training in programming. The training was provided by the vendor.

 In the years that followed, Delta's 1963 system was updated and expanded as more memory and disk drives were added to the system. In 1969 it was necessary to secure a new system. A third-generation system, which had a fairly comprehensive **operating system**, utilities, and a far more powerful language, was obtained. The system had 128K of core memory, tape and disk drives, a card read/punch unit, and a much faster printer than the one used with the second-generation system.

 Their third-generation system was modified to meet Delta Products' specific needs. However, this system also became obsolete. The system illustrated on page 6 was obtained.

 The system illustrated provides a vast amount of computer power which is not being utilized effectively. Second-generation application programs are still being run on a sophisticated system that is capable of doing far more. The system can be upgraded by adding more memory, faster input/output devices, and more control software.

 Today, for the same number of dollars, a new medium-size system with more memory and a physically smaller CPU could be obtainable. It would take less energy to operate, generate less heat, and make calculations and logical decisions far faster than Delta's present system. Although its CPU and operating system might change, many of the same I/O devices could be used.

The CPU's memory consists of 524,288 (1/2 megabyte) bytes of **integrated monolithic memory**. It is possible to expand its memory to a full megabyte. Monolithic memory provides faster storage speeds than magnetic core memory; servicing memory is also easier. The storage cells are implemented on chips that are arranged on cards which can easily be replaced. One character or two digits can be stored in each 8-bit **byte**.

DELTA PRODUCTS' HARDWARE

CPU

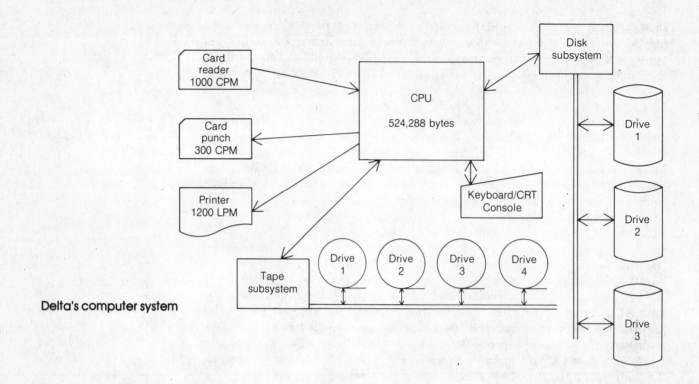

Delta's computer system

Card Reader   The card reader can read 1000 cards per minute. Each of its two stackers can hold 1750 cards. When the first stacker becomes full, the cards are automatically stacked in the second stacker. If the card fails to feed on the first try, three feed retries are made before operator intervention is required. **Photoelectric read** heads are used to read the cards serially by column. The method used for checking the read operation is much improved over former methods. The programmer can tell the system to ignore columns that contain perforations.

Optional features not selected by Delta Products provide for optical mark reading and the two-line multiple-line card print feature. When the mark-reading option is available, the marks are made with an ordinary pencil. The print feature provides for 64 characters per line.

Card Punch   Cards are punched row by row at the rate of 300 cards per minute. Automatic punch retry is provided, and defective cards are sent to a separate error stacker. A two- or multi-line print feature is optional. Delta Products did not elect to have the print feature installed.

Printer   Since a 48-character set is used, 1200 lines per minute can be printed in a **burst mode**. The printer has a 132-print-position line, and either 6 or 8 vertical lines per inch can be printed. As standard features the printer has a powered form stacker, tapeless carriage control, and a built-in vacuum-cleaning system.

Although only four tape drives are available, up to four more can be added before another tape **control unit** is needed. The control unit is a separate piece of hardware that helps coordinate the movement of the data between the tape drive and the memory of the CPU. The data is recorded on 9-track tape at 1600 **BPI**. The tape is read at 100 inches per second (**IPS**) and has .6 inch **IBGs** (interblock gaps). Each standard reel holds 2400 feet of tape. Drives that record 6250 rather than 1600 BPI could have been selected.

Magnetic Tape Subsystem

From two to eight disk drives can be used with a **controller**, which is physically located on one of the drives. Additional drives can be added in one- or two-drive modules. A sealed **data module** contains the disks, read/write heads, and access arms. Each module holds approximately 35 million **bytes** and has two disks. Three of the four surfaces are used for data, and the fourth is used for information needed by the system.

Disk Subsystems

When the system was selected, a great deal of time was spent determining what features should be incorporated into the operating system. The selection committee felt that a VS (**virtual storage**) system would increase the throughput enough to warrant the additional expenditure. The illustration below shows the allocation of memory between partitions storing programs to be executed in a **virtual mode** and partitions storing programs to be executed in a **real mode**.

DELTA'S OPERATING SYSTEM

DOS/VS

     Although there are minimum sizes for the various partitions, the size of each partition is determined when the system is generated. To allow an operating system to be tailored to the needs of the user, a **SYSGEN** (system generation) is done. At this time, the size of the partitions is determined, the system is told what I/O devices will be available, and the desired features are added to the system. When major modifications are made to the operating system, it is necessary to do another SYSGEN. Often this is done when the vendor offers new features that users decide to add to their system.

Shared virtual area
Foreground 1–virtual
Foreground 2–virtual
Foreground 3–virtual
Foreground 4–virtual
Background–virtual
Foreground 2–real
Foreground 1–real
Unallocated real storage for execution of program run in a real mode
Supervisor

Virtual memory jobs stored in F1, F2, F3, F4, and background will run in a virtual mode

Real memory jobs stored in F1 and F2 will run in a real mode

Memory map showing allocation of memory in Delta Products' DOS/VS system

Delta's **DOS/VS** system allows a maximum of five jobs to run concurrently. Programs can run in either a real or a virtual mode. Small programs not well-suited to **paging** should run in a real mode. Paging is a term used to denote the swapping of pages of commands or data to and from real memory. Large programs, usually better suited to paging, should run in a virtual mode. In a real mode, the entire program is within the CPU, and paging is not necessary.

The system manages the portion of real storage where programs from virtual partitions are placed for execution. DOS/VS exchanges sections of programs between the page data set, stored on disk, and real storage (in main memory).

Since jobs running concurrently share the CPU's resources, an internal priority system is used. The **supervisor**, which manages all the system resources, is given the highest priority. Next in order of priorities, from low to high, are background, F4, F3, F2, and F1.

Delta Products' data processing manager was told that programs with a modular structure are best suited for virtual processing. Before the new computer system was used, Delta's programmers seldom used a modular, or **structured approach**, to programming. After the system was installed, inservice training was provided to explain the new programming standards. Many of the concepts related to **structured design** and **structured programming** were included in the standards developed.

POWER/VS

An optional package, POWER, was added to the system to compensate for the fact that data can be processed within the CPU faster than it can be read from punched cards, punched into cards, or outputted to the printer. Under **POWER/VS**, jobs requiring punched card input, punched card output, or printed output are executed in three phases.

1. The job stream (job control language statements and data cards) for various partitions are read in and stored in read **queues**.

2. According to a predetermined priority, jobs are loaded from the queue into the desired partitions and executed. Output to be punched or printed is stored in a punch or print queue.

3. After a job has ended, the output stored in the print or punch queue is printed or punched.

POWER/VS provides additional flexibility. Additional commands, called job execution control language, or **JECL**, permit the operator to delay printing, to print only a portion of a large report, and to reassign data to be punched or printed to some other medium. The job priority, which determines when the job will be executed, is indicated on a JECL card.

Libraries

DOS/VS contains the library programs that maintain and service the libraries. By using the correct job control language statement, programs and source statements can be catalogued into, or deleted from, the libraries. Elements can be copied from one library to another, and the size and location of the libraries can be changed. A library directory can be printed.

LANGUAGES

When Delta Products acquired its third-generation system, BAL (Basic Assembler Language), FORTRAN, PL/I, RPG, and COBOL compilers were supplied by the vendor. Unfortunately, at Delta each programmer had been in the habit of choosing the language he or she wanted to use for a particular application. When the new system was being considered, Delta faced a major decision. Either the data processing personnel would have to rewrite a large number of their existing programs, or they would need to obtain all the compilers they had previously used. Although the compilers were no longer supplied without charge, they elected to obtain FORTRAN, RPG II, COBOL, PL/I, and BAL.

It was their intent to convert existing software to the new system as rapidly as possible so that the old third-generation computer could be phased out. After all systems were operational, they would then consider the acquisition of **CICS** (Customer Information Control System), a software package that supports **online processing** of data and a **database.**

CONVERSION TO AN ONLINE SYSTEM

Once the additional software was added to the system, the Delta programmers planned to study each system to determine how it could be redesigned. They wanted to take advantage of the newer technological advances such as online processing and **distributed data entry**. Before any system would be changed, a complete and comprehensive systems study would be undertaken. A priority system would be developed to determine which system should be studied first.

The redesign of many of the systems was long past due. Most of the system had not been designed for computers. The punched card jobs were converted (rapidly and without being redesigned) to the second-generation computer. When the third generation hardware was installed, the software from the old computer was **emulated** on the new system. Once again when the new system was installed, the old software was converted without being redesigned. The manager was convinced that horse-and-buggy software was being run on a jet-age computer.

Many of the existing systems were still card-oriented. The internal auditors were concerned about the lack of internal and external controls. The accountants often had to search for hours for errors that had gone undetected until reports would not balance. Managers complained that their reports did not provide the information needed. The operations manager was constantly referring to the high number of data exceptions that occurred. Betty Nichols wanted many of the programs modified to provide additional editing.

THE DATA PROCESSING DEPARTMENT

The illustration on page 10 shows the organizational chart for Delta Products' data processing department and lists the personnel who staff it. Although not indicated on the chart, the data processing manager is also the systems manager. The title programmer/analyst is used because Delta favors a team approach that involves the programmers in both the design and implementation of systems. One group of programmer/analysts does both the developmental and the maintenance programming.

The operations supervisor does the tasks usually assigned to the data control clerk and scheduler. The computer operators perform the functions normally assigned to librarians.

If anything, the department is understaffed. Perhaps this is why fire fighting has continued to occur long after most computer service departments have settled into a more relaxed and productive schedule. It seems as if there is never any time to design new systems. Almost 100 percent of the programmer/analysts' time is devoted to the maintenance of existing systems.

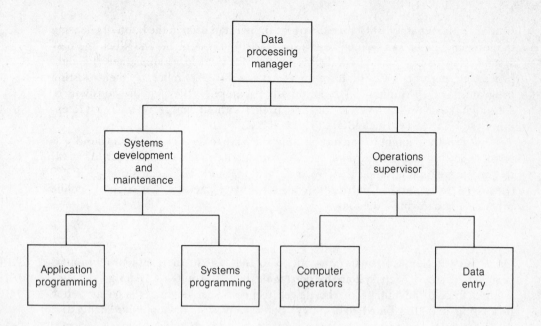

Delta's data processing department

The Staff

Ben T. Paul	Data processing manager
Betty J. Nichols	Operations supervisor
Ron J. Gorney	Systems programmer
Kay Walczak Dennis McNeal Martha Aguilar Mike Arnold	Programmer/analysts
Sue Smith James Long	Computer operators
Cindy Page Peter Gomez Nancy Green Jan Gorney	Data entry clerks
Kim Mead	Programmer trainee

Ron Gorney is primarily concerned with maintaining the systems software, doing SYSGENs, and adding new features to the operating system. He has also written a number of assembler subroutines that have been incorporated into numerous application programs.

The four programmer/analysts spend most of their time maintaining the application software and have very little time to devote to developing new systems. The data entry clerks do many additional tasks—sorting and interpreting cards, disposing of reports, decollating reports, and numerous other tasks usually associated with the preparation of data and distribution of reports.

The manager, Ben Paul, has been with the company for a very short time. He was hired just prior to the selection of the new computer system. Ben Paul felt there was an opportunity at Delta Products to expand the department and to develop new systems. The management of Delta Products felt fortunate to obtain the services of Mr. Paul since he had a good deal of experience in developing online systems. He worked with a VS system that had a CICS software package.

Mr. Paul's former company stressed the use of standards, internal controls, and documentation. His degree is in business data processing and he has worked in a number of supervisory positions.

The previous manager retired after thirty years of service to Delta Products. He had been an accountant until Delta Products acquired punched card equipment. Since he expressed an interest in the equipment, he was asked to develop a data processing department. Each time new equipment was acquired, he attended some type of vendor's school.

The employees within the data processing department were very loyal to the former manager. They enjoyed the relaxed conditions under which they worked. The users felt the former manager did a good job designing their systems. When problems occurred, most users felt they were probably hardware related rather than software related. In reflecting back over the thirty years with Delta Products, the former manager felt the switch from accounting to data processing had been very rewarding.

The new manager, Ben Paul, is concerned about a number of things. Each day brings to light additional problems that need to be solved.

MAJOR CONCERNS

There are many positive things that can be said about the department, but some of the major concerns are:

1. Although there are standards manuals for other departments, none exist for data processing.

2. The documentation consists of little more than source listings, obsolete run sheets, and a few handwritten notes.

3. Although there is a schedule board that shows when jobs are to be run, the operators seldom follow the schedule since there is a constant demand for got-to-have-it-now reports.

4. A number of people have indicated they want additional computer services, but there doesn't seem to be a systematic way of determining priorities.

5. Far too often jobs cancel because of data exceptions.

6. Second-generation application programs are being run on the new system.

7. Not enough internal or external controls were built into most of the systems.

8. Management and the end users have not been directly involved in the design of their systems.

9. Objectives are rarely established for new systems.

Ben feels the staff is competent but in need of more positive direction. The operator who works second shift gets less **throughput** than the first shift operator. Ben also feels that the staff is not aware of current developments in data processing. When he uses terms like CICS and MIS, his staff does not understand the jargon. He feels this is a minor problem since in-service training can be provided.

Ben Paul feels achieving the goals established for the department will be a challenge.

THOUGHT STARTERS

1. Mr. Paul, the data processing manager, is concerned about a number of factors. For each of the following concerns, indicate why you think the problem exists and what you would do (as manager) to solve the problem.

   A. Lack of standards for the data processing department

   Cause _____

   _____

   _____

   Solution _____

   _____

   _____

   B. Lack of good documentation for application programs

   Cause _____

   _____

   Solution _____

   _____

   _____

   C. Posted schedule not being followed

   Cause _____

   _____

   Solution _____

   _____

   _____

   D. Jobs cancel because of data exceptions

   Cause _____

   _____

   _____

   Solution _____

   _____

   E. Lack of internal and external controls

   Cause _____

   _____

   _____

13

Solution _____

_____

_____

F.   Lack of direct involvement of managers and users in the design and implementation of systems.

Cause _____

_____

_____

Solution _____

_____

_____

2.   In looking over your answers to question 1, what one factor do you feel created most of the problems that Mr. Paul has identified?

_____

_____

3.   Will the acquisition of the new computer system and CICS software solve the problems identified?

_____

_____

4.   Would you recommend that Mr. Paul delay the acquisition of CICS software and the development of online systems until some of the problems are solved?

_____

_____

_____

5.   What might have occurred if Mr. Paul, shortly after his arrival at Delta Products, had made the following remark at one of his staff meetings.

Things are sure going to change now that I am manager. Apparently your former manager just didn't understand electronic data processing or how to manage a department. The first thing I will do is to develop standards for developing programs that must be followed by all programmers. The application software that we now have is of a very poor quality.

_____

_____

_____

_____

Name _____

Class _____ Section _____

6. What basic factors concerning motivation and interpersonal relationships would Mr. Paul have overlooked if he had made such a remark?

_____
_____
_____
_____
_____
_____

7. Since there are several different compilers available for the new system, would you let each person code in whatever language he or she preferred?

_____
_____
_____
_____
_____
_____

8. In looking over the material describing Delta Products, would you recommend any changes in either the corporate organization or in the organization of the data processing department?

_____
_____

9. If you were the analyst assigned to a project that would result in the development of a new payroll system, would you feel constraints were imposed if you could not obtain additional hardware or software?

_____
_____

10. Would you recommend that your programmer attend a CICS school (provided by the vendor) prior to the arrival of the additional software?

_____
_____

11. Do the information needs of the various levels of management appear to have been determined and met?

_____

_____

_____

12. Why do you think the consultant recommended that someone other than one of the five founders of the company assume the position of president of the company?

_____

_____

_____

13. If other departments have standards and procedural manuals, why wasn't one developed for data processing?

_____

_____

_____

_____

14. Since Mr. Paul had a good position as manager of a well-established data processing department with another company, do you feel he was unwise in accepting the position offered by Delta Products?

_____

_____

_____

_____

2

DELTA'S PAYROLL SYSTEM

1. To obtain background information on Delta Products' payroll department.   **OBJECTIVES**
2. To obtain an overview of Delta's payroll system.

If you are unfamiliar with any of the following terms or phrases, look them up in the   **KEY WORDS**
glossary of your text or in a data processing dictionary.

Abend reports
Audit trail
Controls
Cost center
Documentation
External controls
Income tax credit
Internal controls
Master file
MBO—Management by Objectives
Post-control sheets
Reconciliation of payroll reports
Timesharing

Before undertaking the tasks assigned, complete the following reading assignment   **PREPARATION**
(or its equivalent): Chapter 3 (The Preliminary Investigation) of *Systems Analysis
and Design*.

TASKS

1. Read the information that follows.
2. Complete the Request for Systems Analysis form on page 22.
3. After you have studied the material, answer the questions on pages 23–25.
4. Review the list of key words.

THE PAYROLL DEPARTMENT

In referring to the organizational chart on page 3, you will find that the accounting department is under the jurisdiction of the vice-president of finance. One of the major divisions within the accounting department is the payroll department. The reporting structure of the payroll department is shown in the illustration on page 19.

In many respects, John P. Schaffer, the vice-president, is an excellent administrator. He has been with the company since it was first organized and takes a great deal of pride in the company. Since he received his formal training a long time ago, his background in electronic data processing is limited. However, he is supportive of most proposals that can be cost-justified. Since 2 percent of the company's gross revenue is spent on electronic data processing, he wants the equipment to be used effectively. It may be because he feels his background in EDP is limited that he leaves most of the decisions regarding EDP to Mary Smith.

Mary Smith has been with the company for ten years. Before joining Delta Product's staff, she was employed as an auditor by a large CPA firm. She is a CPA, and the last four or five years has taken a number of data processing courses at a local university. She feels it is essential that she understands how Delta's data is being processed electronically.

Since Mary has become more knowledgeable in EDP, she is concerned about the lack of **documentation**. In her opinion, the **audit trail** is not as clearly defined as it should be, and there also seems to be a lack of **internal** and **external controls**. She feels this problem exists because the original accounting systems were not designed for EDP. Therefore the capability of the computer to edit and verify data was not used.

The following is a list of specific concerns Mary Smith has about the payroll system.

1. Charges are frequently made to the wrong **cost center**.
2. The **post-control sheets** show that checks are frequently voided.
3. The **abend reports** show that jobs frequently have to be restarted because of data exceptions.
4. Chang spends a great deal of time **reconciling** the various **payroll reports**.
5. When a new report is needed because of a new governmental regulation, the required data never seems to be in the **master file**.
6. Chang produces several manual reports that Mary Smith feels should be generated by the computerized system.
7. Whenever it is necessary to hire a new payroll clerk, a great deal of time is needed to train the new employee. The documentation is of little help since it rarely reflects the current status of a procedure.
8. Much of the data in the payroll master file is also duplicated in the personnel master file. It seems like an excessive amount of time is needed to maintain the files.

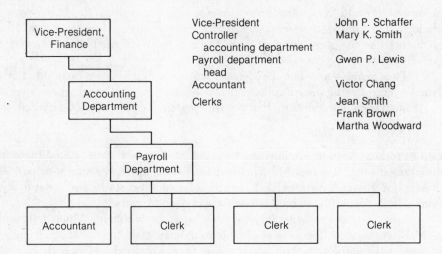

Vice-President	John P. Schaffer
Controller	Mary K. Smith
accounting department	
Payroll department	Gwen P. Lewis
head	
Accountant	Victor Chang
Clerks	Jean Smith
	Frank Brown
	Martha Woodward

Delta Products' payroll department

Although Mary has been concerned about the payroll system for some time, she has made no effort to have it modified. Other items always seemed to have higher priority. Mary had hoped that Chang would take a more active interest in how the payroll data was processed by the computer.

Victor Chang reports directly to Mary Smith and is responsible for the payroll system. Although he has been with the company for five years, Chang still talks about when "he was at General Motors." He was really impressed by their online payroll system. However, when questioned by other members of the payroll department, Chang is unable to answer their questions. Although he knows the EDP buzz words, he has a very limited EDP background.

Each time Smith and Chang met to determine his objectives for the next year (for Delta Products' **MBO** system), Mary indicated he should obtain additional background in EDP. Chang has always had a reason why he could not take a university course or complete an EDP self-study program.

The payroll clerks feel it is difficult to work with Chang. Although Smith, Brown, and Woodward do not report directly to him, Chang treats them as subordinates. Smith and Brown have been with the company four years; Woodward has been with Delta Products for two years.

Mary Smith feels that the clerks are capable and should be given more responsibility. Perhaps many of the errors could be eliminated if the clerks entered the payroll data directly into the system.

THE HISTORY OF THE PAYROLL SYSTEM

Payroll was one of the first applications assigned to the punched card equipment. An analyst designed the system. When the system was first installed, payroll personnel from other companies were called in to observe the payroll data being processed by the punched card equipment. The system was looked upon as a model system, and the data processing manager helped a number of companies install similar systems.

When Delta Products obtained their first computer, the punched card system was computerized. Few other changes were made because shortly after the computer system was installed the board of directors wanted a full report on how the system was working.

When their third-generation system arrived, the programs were recompiled and modified to run on the new system. Since it was considered important to phase out the old computer system as rapidly as possible, the conversion was made without doing a system study. The same controls were used, and the accountants got the same kind of reports they received when the punched card system was used.

Two years ago when Delta's present computer system arrived, the programs were once again compiled on the new system. Although the system now has **timesharing** capabilities, the payroll system is still a card-oriented batch system.

A NEED FOR CHANGE: NEW CITY, STATE, AND FEDERAL REGULATIONS

Delta Products pays its employees bi-weekly, and all of the associated records are maintained on that basis. Management became concerned when it was notified that major changes were needed because of new city, state, and federal regulations. All the regulations would become effective in January.

For the first time city income taxes must be withheld, which will involve several new reports. The new state unemployment laws require that all data pertaining to the employees' hours and wages be maintained on a weekly basis. The new federal regulations make it necessary to identify those individuals who, because of income status, qualify for a weekly **income tax credit**.

In May when Mary Smith read about the changes, she was concerned. Major changes would be needed in the payroll system. Shortly after she read the new regulations, she talked about the changes to Ben Paul.

Although he had not discussed the problem with the payroll department, Ben Paul shared Mary's concern about the payroll system. As they talked over coffee, Ben suggested it might be a good time to do a payroll study. Certainly major changes would be required to comply with the new regulations. This might be an ideal time to request a systems study.

A REQUEST FOR A SYSTEMS STUDY IS MADE

After talking to Ben Paul, Mary decided to talk to Chang. He was predictably unenthusiastic about having a new payroll system. Mary felt he enjoyed finding out why reports were out of balance. Chang indicated it would be difficult to retrain the payroll clerks. As he put it, "they have very limited backgrounds." It is difficult for Mary to understand why Chang resists change.

After her talk with Chang, Mary made an appointment with John Schaffer. Schaffer was impressed with the way Mary presented the information regarding the changes needed. It was one of the few times he completely understood what was said about a computerized system. Mary convinced Schaffer that a new payroll system designed for Delta Products' present computer system would help to eliminate many of the problems. Schaffer suggested that Mary submit a formal request for a systems study.

Mary secured a request form and submitted it to Ben Paul. Without the full support of Chang she requested a systems study to design a new payroll system. She attached to the request form the memo reproduced on the facing page.

Office Memo

TO: Ben Paul, Data Processing Manager   DATE: May 17, 1981

FROM: Mary K. Smith, Controller

RE: New Payroll System

Attached is a request form for a system study to design a new payroll system. Since changes must be made to meet the new regulations, I suggest that this project be given high priority.

I know that you share some of the concerns I have regarding our accounting system. The payroll system is more difficult to work with than any of the other systems. In contrasting Delta Products' payroll system with others that are run on the same kind of equipment, we seem to get less information and have more problems than other companies.

Several supportive documents are attached. If you need additional information, please let me know.

Attachments

A request is made for a systems study

d delta products   REQUEST FOR SYSTEMS ANALYSIS

Section I

DATE SUBMITTED: _____   REQUEST FOR: ☐ MODIFICATION OF SYSTEM

REQUEST IS: ☐ IMMEDIATE   ☐ SHORT-RANGE   ☐ REDESIGN OF SYSTEM

☐ INTERMEDIATE   ☐ LONG-RANGE   ☐ NEW SYSTEM

SUBMITTED BY: _____   _____

(Name)   (Department)

NATURE OF REQUEST: _____

_____

_____

_____

REASONS FOR MAKING REQUEST: _____

_____

_____

_____

_____

SUPPORTING DOCUMENTS ATTACHED: _____

DIRECTIONS: Complete Section I and submit it to the data processing manager. Please attach supportive documents such as new federal regulations, changes in corporate policy, or abend reports. If a new system is being requested, attach any additional information you might have that will help to explain the project.

Section II (to be completed by the data processing manager or an analyst)

MODIFICATIONS APPEAR TO BE: ☐ MINOR   ☐ MAJOR   ☐ EXTENSIVE

IMPLEMENTATION MAY REQUIRE ADDITIONAL: ☐ SOFTWARE   ☐ HARDWARE   ☐ PERSONNEL

COMMITMENT OF RESOURCES WOULD BE: ☐ MINOR   ☐ MAJOR   ☐ EXTENSIVE

INITIAL INVESTIGATION COMPLETED BY: _____

(Name)

PROJECT NUMBER: _____   _____

(Date)

Section III (to be completed after the committee has determined the priority and disposition of the request)

PRIORITY ☐

☐ PRELIMINARY INVESTIGATION
APPROVED

Assigned to: _____

Tentative starting date: _____

☐ UNSCHEDULED BECAUSE _____

_____

_____

DATE REVIEWED: _____

Form No.: DP001
10/15/79

Request form

THOUGHT STARTERS

1. Based upon the material presented in the case study, what documents should have been attached to the request form?

   _____

   _____

   _____

   _____

   _____

2. Assume that you are Kay Walczak and have been assigned to do the preliminary investigation.

   A. Whom would you interview? _____

      Why? _____

      _____

      _____

      _____

   B. How would you make the arrangements for an appointment?

      _____

      _____

      _____

   C. Write a memo confirming the appointment. In the memo state the objectives for the interview.

   D. List at least five questions that you would ask the person you will interview.

      1. _____

         _____

      2. _____

         _____

      3. _____

         _____

      4. _____

         _____

      5. _____

         _____

   E. As soon as possible after the interview, what would you do?

     1. _____

       _____

     2. _____

       _____

3. Assume that you decide to interview both Chang and Smith.

   A. Would you ask them the same questions?

_____

_____

_____

   B. What might be some of the reasons Chang resists change?

_____

_____

_____

_____

_____

_____

_____

4. During the preliminary investigation, what observations might you make?

_____

_____

_____

_____

5. Besides the documents that were attached to the request form, what documents might you examine?

_____

_____

_____

_____

_____

Name _____

Class _____ Section _____

6. If you were on the computer policy committee that reviewed the request, what priority would you assign to the preliminary study? Why?

   _____

   _____

   _____

7. Could Chang's attitude present a problem?

   _____

   _____

   _____

3

PLANNING THE DETAILED INVESTIGATION

1. To obtain more specific information about Delta Products' payroll system.
2. To obtain more specific information about federal withholding tax laws, state and city income tax laws, and social security regulations.
3. To determine the general specifications for a new payroll system.

KEY WORDS

If you are unfamiliar with any of the following terms or phrases, look them up in the glossary of your text or in a data processing dictionary.

Audit	Key field
Backup file	Payroll register
Credit union	Post-run sheet
Detailed investigation	Sequential file
Editing	Source documents
Gantt chart	Spread sheet
General design phase	Stock deductions
General fund check	Terminal
General ledger	Walkthrough
I-S file	W-2 forms

PREPARATION

Before undertaking the tasks assigned, complete the following reading assignment (or its equivalent): Chapter 4 (Detailed Investigations) and Chapter 5 (General Design Specifications) of *Systems Analysis and Design*.

TASKS
1. Read the information in this section.
2. Prepare a Gantt chart for the detailed study that shows who would do each task, the time needed to complete each task (when it would begin and end), and what tasks can be done concurrently.
3. Answer the questions on pages 39–42.
4. Obtain current information regarding:
   a. the percentage method of computing the federal withholding tax;
   b. the current law governing the withholding and payment of FICA (social security tax);
   c. the withholding of income tax for your state.

Information can be obtained from the current *Federal Tax Course*, published by Commerce Clearing House, Inc., or from *Payroll Report Bulletins*, published by Prentice-Hall. Your instructor may have material in your reference library.

BACKGROUND

When the computer policy committee reviewed the request for a systems study, they assigned a priority of one to the development of a payroll study. Since the new city, state, and federal regulations had to be implemented by January 1, Mike Arnold and Kay Walczak were assigned to the project and asked to:

1. conduct an in-depth study to investigate the cause of the problems;
2. complete their investigation by July 31;
3. determine what tasks could be distributed and completed online; and
4. develop the general specifications for the new system.

The new system must be able to run on Delta's present computer system. Because the project team was familiar with the computer system, the **detailed investigation** was to be combined with the **general design** phase. Since Arnold and Walczak had done maintenance on the payroll system, they were familiar with some of the payroll procedures.

Arnold and Walczak started their study by reviewing the preliminary investigation report. Next they brainstormed various ways to complete the detailed investigation. They had to decide:

1. whom to interview;
2. what documents to study; and
3. the best way to study the existing system.

PLANNING THE STUDY

Arnold and Walczak decided they would complete the following tasks in the order listed below.

1. Study the documentation for the payroll system.
2. Study the regulations regarding the new federal, state, and city laws.
3. Review the requirements of the workmen's compensation and social security laws.
4. Study the provisions of the union contracts covering the hourly employees.

© 1981 SRA

5. Review the present system with John Schaffer and Delta's attorney to determine if the present procedures are in compliance with the city, state, and federal regulations.

6. Interview the following people for the reasons stated.

John Schaffer	Determine if the present system provides the information needed by management.
Mary Smith	Determine what additional information the accounting department needs. In addition, find out what controls the department uses to determine the accuracy of payroll information. Determine if more controls need to be built into the computerized procedures.
Victor Chang	Determine the procedures used to verify the payroll data.
Jean Smith Frank Brown Martha Woodward	Determine the format of the payroll data when it is received from other departments; what problems have been encountered, and how the tasks performed by the payroll clerks could be improved.
Betty Nichols	Identify the problems associated with preparing the input and running the payroll programs.

7. Analyze the data collected to determine what:
   a. reports should be changed, deleted, or added;
   b. how, and where, the data should be inputted into the system;
   c. procedures should be changed, deleted, or added;
   d. procedures are needed to create, update, and protect the master files; and
   e. additional controls should be built into the system.

8. Discuss the proposed changes with the individuals interviewed.

9. Incorporate the suggestions of the users into the proposed system.

10. Draft the recommendations and the report.

11. **Walkthrough** the proposed system with the data processing review team (Paul, Nichols, Gorney, McNeal).

12. Prepare the documentation and the final report.

13. Make the oral presentation to the computer policy committee.

A **Gantt chart** was prepared showing who would do each task, the time needed to complete each task, when each task would be completed, and what tasks could be done concurrently.

PROCEDURES RUN UNDER THE PRESENT SYSTEM

The procedures within the present system are either classified as display, maintenance, or report programs. Each procedure (or program) is briefly explained and listed in the order performed. The present system is a bi-weekly system. Management has indicated the new system must also be bi-weekly.

Display Program

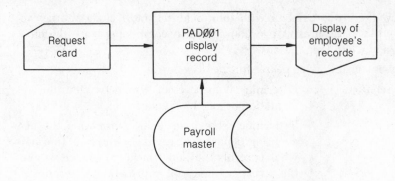

A card containing the employee's name is used to retrieve the employee's record from the master file. The **key field** within the records stored in the **indexed-sequential** payroll master file is the employee's name. The employee's name is used as the key because John Schaffer wanted some of the reports printed in alphabetical order by the names of the employees. Since the current file is also in sequence by employee name, it sometimes presents a problem when employees change their names.

Members of the payroll or data processing department request the program run to determine the data stored in an employee's master record. Some fields of data recently added to the master file records cannot be displayed.

Utility disk-to-printer programs are available. However, PADØØ1 formats the data while the utility programs do not.

Maintenance Programs

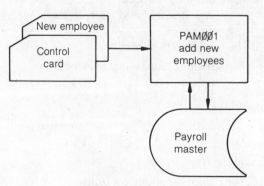

PAMØØ1 adds new employees to the indexed-sequential master file. Three cards of data are needed to create an employee's master file record. In examining the program, it was found that the data coming in from the card was not edited. Usually when an **audit** is done on the **payroll register**, errors made in creating an employee's master file record are detected. Since the errors are not detected until then, the master file records with wrong data need to be corrected and the payroll register program rerun. The analyst who designed the system felt it was not necessary to print a report listing the employees added to the file.

PAMØØ2 is used to make changes and corrections to the data stored in the master file records. Any field within any record can be changed. Two types of changes are made. The first type is what might be considered a normal change—an employee moves, has a change in salary, or wishes to change his or her **credit union** deduction. The second type is a change needed to correct a past

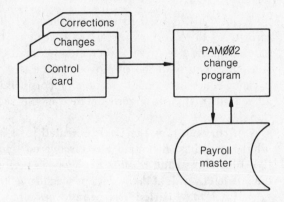

error. For example, a check was issued to John Green by mistake. Although John had not worked, his supervisor submitted a time sheet.

Both the data entry and payroll clerks feel the source documents are very confusing. Since several card formats are used, it is difficult to transfer data from the source documents to the punched cards. Because of a keypunching error or an error in recording the data on the document, incorrect changes are sometimes made.

Very little editing is done, and errors are often undetected until the payroll register audit. Chang usually finds that some authorized changes were not made. If the corrections are not made properly, reports get out of balance. When that occurs, it is difficult to determine what changes were made because the source documents are not retained. Since the systems analyst felt it would not be useful, a report for the payroll department is not printed.

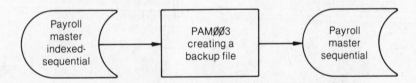

PAM003 is used to create a sequential payroll master file from the indexed-sequential payroll master file. The file is copied without changing any data. The sequential file serves as a **backup** for the master file and can be sorted into any desired sequence.

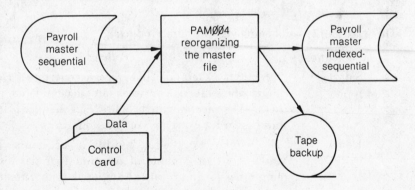

PAM004 is used to recreate the payroll master file. When this is done, records in the overflow area are placed in their proper sequence. When the payroll master file is being recreated, records can be deleted. If a record is to be deleted, a card with the employee's name is required. The control card indicates if changes are to be made to other records. For example, it is possible to delete all records that have a zero balance in the year-to-date salary and bond deduction fields.

A backup tape file is also created for storage in a fire-proof vault located off the premises. Ben Paul has discovered that the person assigned to take the tape files to the vault is often negligent in performing this task. The tapes are frequently left overnight in the data processing center.

Each pay period new employees are added to the file, and then the necessary changes and corrections to the master file are made. PAM003 is run to provide backup for the updated file. After the master file is updated with the new payroll data created during the payroll register program, PAM003 is run again. Immediately after running PAM003, PAM004 is run. Twice an operator ran PAM004 prior to running PAM003, and the obsolete year-to-date information replaced the updated information. Betty Nichols has requested that PAM004 be modified so it would be impossible to run it before PAM003 is run.

Report Program

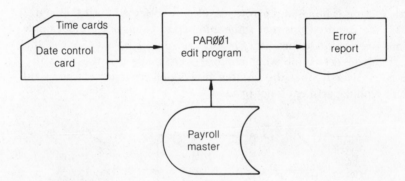

PAR001 edits and checks the sequence of cards. If an employee works for more than one cost center, a card is used for each center. The program also checks a number of different factors such as the validity of the account and cost center numbers. If errors are detected, the program is rerun until a "no errors in run" message is printed. In order to be paid, an employee must have one or more time cards. Ben Paul feels it is unnecessary for all employees to have time cards.

EMPLOYEE CLASSIFICATIONS

There are four types of employees at Delta Products.

Classification	Policies
Salaried	Salaried workers do not receive overtime. However, all hours above forty per week are accumulated in a special field. The accumulated hours can be used for personal business or for extra days of vacation.
Hourly	The regular work week consists of 35 hours (7 hours per day). The first 3 hours of overtime (per day) is at 1.5 the employees' regular rate. Additional hours are at 2.0 their

regular rate. A maximum of 15 hours of overtime per week is allowed.

Contract   A few employees are classified as contract employees and receive a set amount to be paid over a given period of time. For example a contract programmer might receive $5000 over a ten-week period. The programmer would receive $1000 for each of five pay periods. At the end of that time, no more payments would be made unless a new contract was issued. Contract employees determine their own work schedule.

Part-time   Part-time employees are limited to 20 hours a week and are not paid overtime.

Since the union contract states that only a certain percentage of the total payroll can be paid for contract and part-time employees, it is essential that the status of all employees be correctly determined. The union also checks to see that part-time employees do not work more than 20 hours a week and full-time hourly employees do not work more than 15 hours of overtime a week. The problem is compounded since during the year an employee's status can change. Part-time becomes full-time; hourly becomes salaried; and contract employees sometimes become salaried employees.

PAYROLL REGISTER PROGRAM

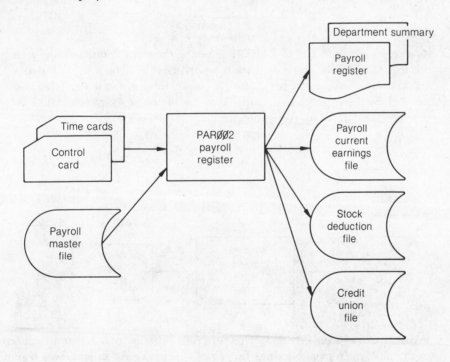

Edited time cards and the payroll master file are used as input to PARØØ2. The payroll register lists each employee's regular and overtime hours, gross earnings, and withholdings for federal tax, state tax, social security, stock, and credit union. Each employee's net pay is also calculated and printed.

The department summary shows how many people in each of the four categories were paid. The report also shows the total gross earnings charged to each department. The accounting department uses these figures to determine the accuracy of the monthly payroll distribution report. The department heads are unaware of the report. The department heads feel there should be some way of determining on a bi-weekly basis if the correct amount is charged to their department.

PARØØ2 is run at least twice. The first time the report is printed on two-part stock paper. Chang and Jean Smith audit the register. Often they find that an employee is being paid for an incorrect number of hours or at an incorrect rate of pay. Sometimes a terminated employee appears on the register. It is often necessary to make changes in the master file before the final run of PARØØ2. During the final run, the register is printed on a better quality two-part paper than the paper used during the edit runs.

STOCK DEDUCTION REPORT

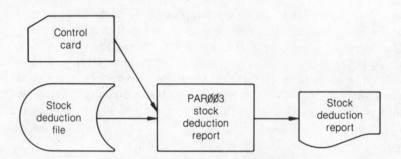

As soon as the final run of PARØØ2 is made, the **stock deduction** file is used as input to PARØØ3. The stock deduction report lists the details of each person's stock account—the balance in his or her account, total amount deducted, how many shares have been purchased, and the value of the stock purchased. The value is based on the current market value.

CREDIT UNION REPORT

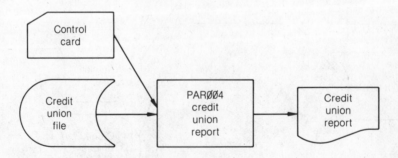

PARØØ4 is run after PARØØ3. This program lists the total amount deducted from each employee's gross earnings for deposit in the credit union. Two copies of the report are printed because one copy is needed to accompany the check that is sent to the credit union. The report shows both the current and the year-to-date deductions.

PRINTING THE PAYCHECKS

The payroll current earnings file is used as input to PARØØ5, which is the program that prints the paychecks. The summary printed at the end of the job lists the totals for the fields printed on the check. The totals are compared with the ones listed on the payroll register.

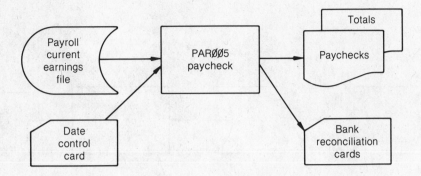

The punched cards are interpreted and used in the program that reconciles the bank statement. A special account is used for the payroll. Each pay period a transfer is made from the general fund to the payroll fund.

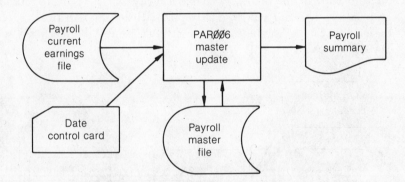

UPDATING THE PAYROLL MASTER FILE

The current values stored in the payroll current earnings file are added to the quarterly and year-to-date totals stored in the employees' master file records. The payroll summary is also printed. The totals are compared to the ones on the payroll **post-run sheet**. Each pay period, the payroll register totals are added to the year-to-date totals. The post-run sheet totals must agree with the payroll summary totals. When corrections are made, entries are made (and documented) on the post-run sheet. Sometimes the totals (the post-run sheet and payroll summary) do not agree. This generally is the result of an incorrect adjustment for a previous error.

MONTHLY PROCEDURES

Payroll Distribution Report

Two months during the year there will be three current files because the employees will receive three instead of two paychecks. The rest of the year only current file 1 and current file 2 are used as input to the sort/merge program. (See page 36.) The distribution report shows the charges made to each account number within each cost center. The additional payroll charges for taxes and FICA to be prorated to each department are also calculated and printed. Since only the totals are printed, the department heads are unable to calculate the accuracy of the report. Chang reconciles the totals with the two (or three) payroll register summary reports. The distribution cards are used in the **general ledger** accounting program to update the general ledger and department accounts.

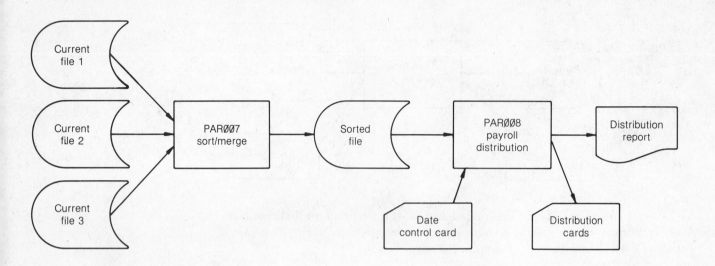

QUARTERLY PROCEDURES

A work copy of the quarterly social security report is printed and audited prior to printing the final report. The report is listed alphabetically, which makes it easier for Chang to audit. After Chang verifies the accuracy of the report, PAR009B is run.

FICA Report

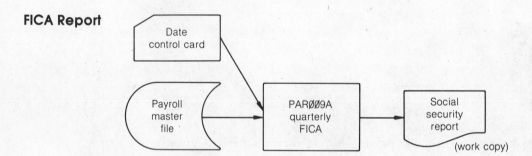

PAR009B is a two-phase program since the sequential version of the payroll master file must first be sorted into sequence by the employee's social security number. The report lists each employee's name, social security number, taxable FICA wages, and the FICA withheld.

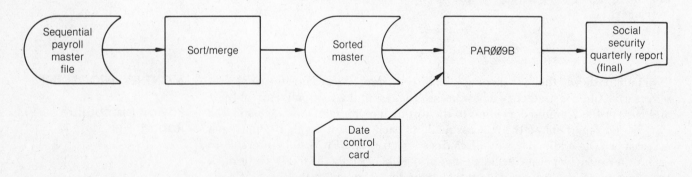

At the top of each page is printed the employee's name, address, and identification number, the page number, and the date the quarter ended.

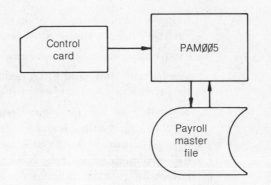

Quarterly Change Program

After the quarterly reports are produced, PAMØØ5, which changes the quarterly totals to zero, is run. At the end of the fourth quarter, the year-to-date totals are also changed to zero. At the end of each quarter, the master file is copied to a quarterly backup tape file.

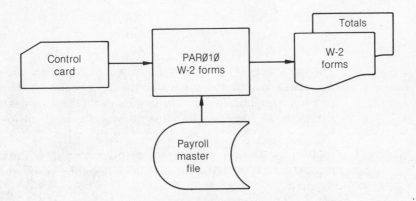

ANNUAL REPORTS

W-2 Forms

After the **W-2 forms** are printed, totals are printed that must agree with the totals printed on the payroll summary. Usually the W-2 forms are printed before the first pay period in January. If the year-to-date summary and the W-2 forms aren't printed prior to that time, an I-S file can be created from the yearly backup file and used as input to PARØ1Ø.

GENERAL COMMENTS ON THE PAYROLL SYSTEM

The following comments were made by the analysts. They were based upon observations or upon information received by interviewing department heads, members of the payroll department, or other members of the data processing staff.

1. A large percentage of the data entering the system is not edited.
2. The system is card-oriented.
3. A different format is used for each date control card.
4. Department heads find it difficult to determine if they are charged for the correct amount of labor. They have no way of checking the accuracy of the monthly payroll distribution report.
5. Although there are only fifteen payroll classifications for the following types (hourly, salaried, and part-time employees), whenever a cost of

living adjustment occurs, or an employee receives a raise, the pay rate in each person's master file must be changed.

6. Frequently jobs cancel because of data exceptions.

7. The federal, state, and FICA deductions seem to be in compliance with the laws. The reports are verified and sent in on time.

8. According to the new state regulations, information regarding the regular and overtime hours for each employee should be maintained on a weekly basis. When an unemployment report needs to be submitted, the accountants must refer back to the timesheets. The required data is not available in the employees' master file records.

9. The weekly tax credit to be paid to employees who qualify should be a part of the payroll system. Separate **general fund checks** should not be issued.

10. The city tax regulation has led to some confusion since the following factors apply:

Percent Paid	Live in City	Work in City
1	yes	yes
0.5	no	yes
0	no	no
1	yes	no

The city income tax question is a little more complex since Delta Products has two locations—one within the city; one outside the city. The city treasurer has indicated that for nonresidents the wages earned working within the city are taxable while the wages earned outside the city are not.

11. Some of the data recorded in the payroll master file is also duplicated in the personnel file. Management feels a more adequate database is required.

12. The union feels they should be given a monthly report that shows both the current amounts paid in each category and the year-to-date totals in each of the four categories. They would like the report printed by cost center and account number.

SUMMARY   Walczak and Arnold completed their study and submitted their recommendations to the computer policy committee. Their report indicated that the present system was inadequate. Although the external reports were accurate and submitted on time, the internal reports were inadequate and sometimes incorrect.

Their report indicated it was essential to develop an integrated database to better meet the needs of the payroll and the personnel department. Department heads and other levels of management must be able to obtain reports that are timely and more complete.

While some of the tasks could be decentralized and performed within the payroll department, most of the procedures would still be executed within the data processing department. Although one of the constraints was to use existing equipment, the analysts recommended that a **terminal** be placed in the payroll department.

THOUGHT STARTERS

1. Each of the procedures fall into one of three general classifications. The three classifications are:

   A. _____

   B. _____

   C. _____

2. Which of the procedures described should be decentralized? Under the concept of distributed data entry, the data would be entered into the system by the payroll clerks.

   A. _____

   B. _____

   C. _____

3. Based upon the information provided, what rationale would you use to justify the distributing of the procedures identified in question 2?

   _____

   _____

   _____

   _____

   _____

   _____

   _____

4. Complete the chart regarding the procedures that make up the present system. Indicate if you would drop the procedure (D), modify the procedure (M), or retain (R) the procedure without any major modifications. If you feel a procedure should be modified, briefly state how you would change the procedure. In completing the chart, under the code column record a D, M, or R.

Procedure	Code	Modification Recommended
PAD∅∅1		
PAM∅∅1		
PAM∅∅2		

Procedure	Code	Modification Recommended
PAM003		
PAM004		
PAR001		
PAR002		
PAR003		
PAR004		
PAR005		
PAR006		
PAR007		
PAR008		
PAR009A		
PAR009B		
PAR010		

Name _____

Class _____ Section _____

5. What procedures would you add? State the rationale for adding the procedure.

_____

_____

_____

_____

_____

_____

_____

_____

_____

_____

6. What changes might be made that would address the problem identified as "Department heads find it difficult to determine if they are charged for the correct amount of labor. . . ." Identify what type of report you would provide and how the procedure would be integrated into the payroll system.

_____

_____

_____

_____

_____

_____

_____

_____

_____

_____

_____

_____

_____

_____

7. Review the general comments on the payroll system. Based on the comments, list five changes you would make to the present system.

_____

_____

_____

_____

_____

_____

_____

_____

_____

4

DETERMINING I/O REQUIREMENTS

1. To obtain more specific information about Delta Products' payroll system.   **OBJECTIVES**
2. To determine the output to be generated from each procedure.
3. To determine the input needed for each procedure.
4. To design the master and transaction files.

If you are unfamiliar with any of the following terms or phrases, look them up in the   **KEY WORDS**
glossary of your text or in a data processing dictionary.

Alphanumerical	Modular
Cost centers	Packed numeric data
Current transaction file	Top-down programming
Data control	Zoned numeric data
Date-control file	

Before undertaking the tasks assigned, complete the following reading assignment   **PREPARATION**
(or its equivalent): Chapter 6 (Detailed Design: Output), Chapter 7 (Detailed Design:
Input), Chapter 8 (Detailed Design: Files), and Chapter 9 (Detailed Design: Determining Procedures) of *Systems Analysis and Design*.

1. Read the information in the section.   **TASKS**
2. Answer the questions on pages 55–57.
3. Using Exhibit 2 and 3 as a guide, list the input and output requirements for each
   procedure. To do this, determine what output is needed for each procedure. Af-

ter this is done, determine how the data is obtained. It could be read from the **date-control file**, a **current transaction file** or the master file. Some fields of data to be printed or written into an output file will need to be calculated.

It is suggested that you start with the payroll register procedure because many of the master file fields will be needed in this procedure. If all the data is in a file which has been defined, you may indicate that the file is read as input, and you need not list each individual field.

It will be impossible to do some of the procedures until after the content of the master file has been determined. Examples are the program that adds new employees to the master file and the program that displays the contents of the master file on the CRT.

You may also find that as you proceed through the assignment, you will need to go back and revise some of the files from which data was omitted.

4. Using the file layout sheets provided, do the file layouts. For each field determine if the data should be **zoned**, **packed**, or in a character format. Also determine the size of each field. Since the current files, other than the current earnings file. will contain relatively short records, more than one file may be shown on a form. However, use a separate sheet for the payroll master file and for the current earnings file.

ADDITIONAL PAYROLL INFORMATION

The following information must be used in developing the payroll system. In some cases the material presented is a brief review of material presented in previous assignments.

1. There are four classifications of employees (see page 32). Each classification should be considered individually in determining the input needed, the editing procedures required, and the methods to calculate gross pay.

2. The union contract is very specific regarding overtime and the hours per week part-time employees may work (see page 32).

3. Deductions are made from the employee's gross pay for the following reasons.

Federal, state, or city regulations

Federal income tax	Based on the computational tax method.
State income tax	In compliance with the state regulations.
City income tax	1% will be deducted if the employee lives in the city; 5% if the employee lives out of the city. Some nonresidents work in both of Delta Products' locations. Only the portion of the employee's gross pay earned while working in the city is taxable.
FICA	All employees are subject to social security tax regulations.

Note: Your instructor may provide the forms needed for completing task 3. If not, rule forms similar to the one on page 52.

Voluntary deductions

Credit union	The amount to be deducted each pay period is stored within the employee's master file record. Whenever they wish, employees may change the amount of their deductions.
Stock	The amount to be deducted each pay period is stored within the employee's master file record. Whenever there is enough money in the employee's account to buy a share of stock, that amount is deducted from the account. The cost of stock to employees is 20% less than the current market value of the stock. The value of the employees' shares is computed at the full market value.
Term insurance	Group term insurance is available to the employees who wish to participate in the plan. The amount to be withheld is in each employee's master file record.
United Fund	Employees may elect to have United Fund contributions deducted from their gross earnings. Each employee elects to contribute a given amount of money, which is deducted over a given period of time. For example, an employee may wish to contribute $200 and have it deducted over five pay periods. All United Fund deductions start with the first pay in January and stop whenever the required number of deductions have been made.

Involuntary deductions

Union dues	All full-time hourly employees pay union dues. The amount of each employee's contribution is in his or her master file record.

4. Delta Products has fifty **cost centers**, numbered from 1–50.

5. Account numbers are numbered from 1–100. The payroll account numbers are 11–29. The account number provides a further breakdown of the four major employee classifications and is used for some of the cost-accounting functions.

6. The monthly payroll distribution report shows the distribution of earnings by account number within each cost center.

7. The cost center and account number that each employee normally works under is stored within his or her master file record. Employees often work under more than one cost center and account number.

8. The amount of the employee's income tax credit is stored in his or her master file record. The amount paid to employees is reflected in their net pay but cannot be included in gross pay. The monthly report submitted to the federal government must show the monthly and year-to-date totals.

Delta Products will be reimbursed for the amount of the tax credit paid to employees. Very few employees qualify; nevertheless, the plan must be available for those who do qualify.

9. One supervisor or manager within each cost center is responsible for verifying the accuracy of the information recorded on the payroll sheets. The payroll sheets are printed after the necessary additions and changes have been made to the master file.

   a. A separate sheet is available for each cost center and lists all employees normally assigned to the center.

   b. Hourly and part-time employees have time cards and use a time clock for recording the hours worked within each center. The regular and overtime hours for hourly and part-time employees must be recorded on the payroll sheets.

   c. Overtime for salaried employees, which is not paid in wages but credited to a compensatory time account, is added to the form.

   d. No entry is needed for contract employees other than to verify that they are to be paid. The names of employees to be paid, but who are not on the form, must be added by the supervisor.

10. The following timetable is used in processing the bi-weekly payroll. Dates are used to illustrate the exact time the various procedures will be implemented.

Wednesday (Oct. 1)	The master file is updated by adding new employees and making other necessary changes. A termination date is entered for any employee who is no longer with the company. The paysheets are printed and distributed to the centers.
Thursday (Oct. 2) 12:00 P.M.	Pay period ends.
Friday (Oct. 3) 4:00 P.M.	Payroll sheets are due from all cost centers.
Monday (Oct. 6)	The payroll sheets are checked by the payroll clerks, and batch totals are established. Totals are accumulated for each cost center. The total number of employees to be paid in each classification is determined. For hourly and part-time employees, totals are established for regular and overtime hours. The amount of compensatory time is totaled for all salaried employees.
Tuesday (Oct. 7) 9:00 A.M.	The payroll sheets and batch total sheets are delivered to **data control**. Data control logs in the forms and submits the forms to data entry.
Tuesday (Oct. 7) 1:00 P.M.	The payroll sheets are returned to data control. The time cards are sequenced on the sorter by employee names. PAR001 is run. The data control clerk checks the error report. The cause of each error listed is determined. In some cases it is necessary to have the payroll clerks submit additional changes to

the master file. PAM002 will need to be re-run. After a clean edit run (one with no error messages) is obtained, PAR002 is run. The data control clerk checks the department summary report with the batch total sheets. Differences must be resolved. It may be necessary to rerun PAR002. If the summary totals agree with the batch total sheets, the forms and the preliminary register are submitted to the payroll department.

Wednesday (Oct. 8) 9:00 A.M.	Necessary changes are submitted to data control. Data entry punches the required cards; new time cards are merged with the time cards used on Tuesday. Additional changes to the master file are made. An edit run is made of PAR002. Data control confirms that all required changes have been made and authorizes the final run of PAR002.
Wednesday (Oct. 8) 1:00 P.M.	Final run of PAR002. Data control posts the totals to the post-run sheets in order to confirm the accuracy of PAR006. PAR003 and PAR004 are run and the validity of the reports are confirmed. The checks are printed by running PAR005. Data control confirms the totals and the number of checks printed. If all of the reports are valid and there are no apparent errors, PAR006 is run to update the master file and to print the payroll summary.
Friday (Oct. 9) 9:00 A.M.	Checks are available for distribution.

Since the timetable allows for contingencies, the checks have always been available by Friday, 9:00 A.M.

11. Since the payroll distribution report is by account number within cost center, hourly and part-time employees may need more than one time card. If an employee works in more than one cost center or has more than one account number, a separate payroll current earnings file must be written for each cost center and account number. Only the last record for an employee will include his or her voluntary and involuntary deductions. Each employee will receive only one paycheck and only one line will be printed on the payroll register.

All of the facts listed under items 1 to 11 must be taken into consideration when designing the new payroll system.

GENERAL GUIDELINES FOR DEVELOPING THE PAYROLL SYSTEM

The following guidelines were included in the report submitted after the detailed investigation and general design phase of the project were completed. Management and the computer policy committee agreed with the suggested guidelines and instructed Walczak and Arnold to incorporate the guidelines into the detailed design of the new payroll system.

1. Although the system to be designed will be less card-oriented, punched cards will still be used for entering data into the edit program. Output created on disk from the clean run of the edit program will be used as input to the payroll register program.

2. An exception system will be developed. Time cards will be used in the following cases.

Classification	Used for
Hourly	Employees who work more or fewer than 35 hours per week; employees who work under more than one account number or cost center.
Part-time	All employees
Salaried	Employees who are not to be paid and those who have compensatory time.
Contract	Deferring a payment until a future pay period.

3. Rather than using employees' names as the key to their master file records, an **alphanumerical** number will be assigned to each employee. The first employee in alphabetical order will be assigned the number 20. The second, 40, and so forth. This will allow for insertions into the file as new employees are hired. Although at the end of the year terminated employees will be deleted from the file, their numbers will not be reassigned. A report (PAR004) will be available upon demand that provides the payroll department with a listing of all current employees. Only employees with a termination date of zero should be listed on the report. The following data will be listed for each employee.

Employee's number	Number of exemptions
Account and cost center	Hourly rate*
Name	Salary*
Social security number	Contract, number of installments, and installments paid*

The report will be used in some of the payroll audit functions. In addition, when a new employee is assigned a number, the payroll clerk will need to reference the report.

4. Since additional programs will be added to the systems, the programs will be assigned new reference numbers.

5. A terminal will be available in the payroll department for use in executing the following programs.

PAD001	Displaying the employee's record. If necessary, the display can be queued to a printer.
PAM001	Establishing a master file record for a new employee.
PAM002	Making corrections to employees' master file records.
PAM003	Making what are considered normal changes to employees' master file records.

In addition, the terminal will be used for initiation of the report-

---

*Data printed only if it applies to the employee.

producing programs that are executed upon demand. Programs that fall within this category are PADØØ1 and PARØØ2.

6. Although the functions listed in number 5 are distributed, the appropriate form must be completed and approved before changes or additions can be made to the payroll master file.

7. Additional reports must be available. For example, when changes are made to the master file, a printout must be available as confirmation of the change.

8. After the edit run of the payroll register program, a detailed report is to be printed for each cost center. Each cost center's report should provide the following information:

Employee's number   Hours—regular and overtime

Employee's name   Gross earnings

Account number

If an employee works under more than one account number, a separate line will be used for each number. Before the final run of PARØØ4, the supervisor who approves the time sheets will audit this report.

9. All files will be written and maintained on disk. Tape will be used to back up the payroll master file and the current files retained beyond the current fiscal period.

10. Rather than a separate card for each date-control record for each run, the data needed for the entire bi-weekly payroll system will be recorded in one disk record. The record will provide the information needed for all bi-weekly payroll programs. The control information will be entered into the system by terminal. Each field will be confirmed after the data is entered.

11. Output files written to produce reports must provide all of the data needed. As the current file is being written as output, master-file information needed for the report will be transferred to the current file.

12. The payroll programs used under the old payroll system will be examined to determine if:
    a. new data entering the system is edited and verified;
    b. the validity of the output is confirmed;
    c. the correct amount of detail is included in printed reports;
    d. a clear audit trail is available; and
    e. the coding conforms to the established standards.

13. Programs that are not **modular** and not written using a **top-down approach**, should be rewritten.

14. The documentation should conform to established standards.

15. The timetable used for the current system will be used when the new system is operational.

16. Suggested changes to existing procedures will be discussed and approved by users and management.

Arnold's and Walczak's report included a summary of the procedures that would be necessary to implement the new payroll system. The summary is shown on the following page. The new reference numbers, such as PAMØØ1, will also be used in cataloging and executing the programs.

EXHIBIT 1: Programs Needed to Implement the New System

Number	Description	D/C*	Input	Output
PAD001	Formatted display of employee's record	D	Payroll master file	Formatted display of employee's record
PAM001	Add new employees to the master file	D	Data from terminal Master file	Updated master file Confirmation report
PAM002	Corrections are made to the master file	D	Data from terminal Master file	Updated master file Confirmation report
PAM003	Changes are made to the master file	D	Data from terminal Master file	Updated master file Confirmation report
PAM004	Copy I-S file to a sequential disk file	C	I-S master file	Sequential master file
PAM005	Recreate I-S master file. Some changes may be made as the new file is written.	C	Sequential master file Data—punched cards	I-S master file Tape backup file
PAM006	Changes quarterly totals to zero. At the end of the fourth quarter, year-to-date totals are also changed to zero.	C	Master file	Updated master file
PAR001	Listing of employees showing their current payroll status	D	Master file	Status report
PAR002	Cost center time sheets	D	Master file	Cost center time sheets
PAR003	Time cards are edited	C	Time cards Master file	Error report Transaction file
PAR004	Payroll register and current files for additional reports are written.	C	Master file Transaction file Rate file to load rate table	Payroll register Current earnings file Stock deduction file Credit union file Tax credit file Weekly earning file

	Description		Files	Reports/Output
PAR005	Cost center reports to be used in auditing the current payroll charges to the center.	C	Weekly earnings file	Cost center audit report
PAR006	Shows the current deductions by employees for stock	C	Stock deduction file	Stock deduction report
PAR007	Credit union report	C	Credit union file	Credit union report
PAR008	The bi-weekly income tax credit report is printed.	C	Income tax credit file	Income tax credit report
PAR009	Prints the paychecks and writes a bank reconciliation file.	C	Current earnings file	Pay checks Bank reconciliation file
PAR010	Prints the weekly time and earning register.	C	Weekly earnings file	Weekly hour and earnings report
PAR011	Updates the payroll master file by adding values from the current earnings file.	C	Master file Current earnings file	Updated master file Listing of values added to the file
PAR012	Prints the year-to-date payroll summary.	C	Master file	Payroll summary report
PAR013 PAR014	The current earning files are merged and sorted to produce a monthly distribution report. A file is also created that will be used to update the general ledger accounts.	C	Current earnings files	Payroll distribution report Distribution file
PAR015 PAR016	The sequential master file is sorted by social security number to print the quarterly social security report.	C	Sorted master file	Social security report
PAR017	The W-2 forms are printed.	C	Payroll master file	W-2 forms

Note: A comprehensive payroll system might include more procedures and reports. However, for the purposes of the case study only the ones listed will be considered. Although not listed, in almost all cases the date-control record is required.

---

*D indicates the function is distributed; C indicates the function is centralized.

EXHIBIT 2: Social Security Report

PROGRAM: PARØ16

FUNCTION: Print W-2 forms

INPUT:
- **X** Payroll master file—sequenced by _____
- _____ Transaction file
- _____ Punched cards
- **X** Date control record

OUTPUT:
- _____ Transaction file
- **X** Printed report
- _____ Updated master file

HEADING 1: CONSTANTS TO BE PRINTED ON EACH FORM: Employee's social security number, employee's name and address, identification code

HEADING 2: _____

TOTALS: _____

INPUT	CALCULATIONS REQUIRED	Z/P/C	SIZE	OUTPUT/NAME OF FIELD	COMMENTS
Payroll master file	Totals for:			Employee's:	
	Federal income tax withheld			Social security number	
	FICA withheld			Employee's name and address	
	State income tax withheld			Federal income tax withheld	
	City income tax withheld			Wages	
	Employee's gross earnings			FICA	
	FICA covered wages			Total covered wages (for FICA)	
				State income tax withheld	
				City income tax withheld	
Date control card				Totals for:	
				Federal income tax withheld	
				FICA withheld	
After W-2s are printed, totals are printed on stock paper				State income tax withheld	
				City income tax withheld	
				Employees' gross earnings	
				FICA covered wages	

© 1981 SRA

EXHIBIT 3: Cost Center Time Sheet

PROGRAM: PARØØ2

INPUT: **X** Payroll master file—sequenced by   Cost center, employee number
       ___ Transaction file   ___ Punched cards   **X** Date control record
       ___ Transaction file   ___ Updated master file

FUNCTION: Print cost center time sheets

OUTPUT: ___ Transaction file   **X** Printed report

HEADING 1: TIME SHEET FOR nnn COST CENTER DATE   nnn is cost center number from MF; print closing payroll date

HEADING 2: EMPLOYEE NUMBER, NAME, CLASSIFICATION, ACCT. NUMBER, NO PAY, REGULAR HOURS, OVERTIME HOURS, COMPENSATORY HOURS

TOTALS: TOTALS EMPLOYEES TO BE PAID nnn

INPUT	CALCULATIONS REQUIRED	Z/P/C	SIZE	OUTPUT/NAME OF FIELD	COMMENTS
Date control record	Accumulate by cost center			Employee number	
	the total number of			Employee name	
Payroll master file	employees				
				Account number	
				Classification	

Additional Instructions: Please print blanks under the captions for no pay, regular hours, overtime hours, and compensatory hours.

Double space the report. Each cost center's report is to be printed on a separate sheet.

Note: For a report, the data representation would be a C. The size of the field cannot be determined until the master file is designed.

THOUGHT STARTERS

1. Of what significance is it that the account numbers range from 1–100 and the payroll numbers range from 11–29?

   _____

   _____

   _____

2. Assume that Charles Wilson normally works under cost center 55 and account number 18. Why would he need a time card if he worked under cost center 65 rather than cost center 55?

   _____

   _____

   What would need to be done if he worked 17 hours under cost center 55 and 18 hours under cost center 65?

   _____

   In the above case, why are two payroll current earnings files written?

   _____

   _____

   _____

3. Under what circumstances will employees have time cards?

   A. _____

   _____

   B. _____

   _____

   C. _____

   _____

   D. _____

   _____

4. Assume that 80 percent of Delta Products' employees work during the first pay period in January. Will the master file be accessed sequentially or randomly?

   _____

   _____

Would you have answered the above question differently if you had been told that normally only 25 percent of the employees work in any given pay period?

_____

_____

5. In what ways is the new system less card-oriented than the old system?

_____

_____

_____

6. Why is it important that the analyst understand the union contract regarding overtime and the employment of part-time employees?

_____

_____

_____

_____

7. Why is it better to use the employee number (an alphanumeric code) for the key field in the master file than the employee name?

_____

_____

8. What technique would you recommend for verifying data entered from a terminal to create a record for a new employee or to correct a field of data in an employee's master record?

_____

_____

_____

9. How would you respond to the data processing manager if he or she asked, "Why are you printing a report listing the data used in creating a record for a new employee? What good is the report?"

_____

_____

_____

_____

10. Indicate why you would either agree or disagree with the recommendation of Walczak and Arnold to send a report to each cost center that listed the employees' earnings that would be charged to the cost center? The super-

Name _____

Class _____ Section _____

visor will be asked to audit the report and notify the payroll department of any errors.

_____

_____

_____

_____

_____

_____

11. What are the advantages of using one date control record for all the payroll programs?

_____

_____

_____

12. Why should the analyst determine the output and input needed for all of the payroll programs before designing the payroll master file?

_____

_____

_____

13. Why is it a good idea to determine what calculations are required in each of the procedures?

_____

_____

14. Do you agree with Walczak and Arnold's recommendation that disk rather than tape files be created? Why do you either agree or disagree with their recommendation?

_____

_____

_____

RECORD LAYOUT AND TEST DATA

RECORD NO.

RECORD DESCRIPTOR

© 1981 SRA

RECORD LAYOUT AND TEST DATA

RECORD NO.

RECORD DESCRIPTOR

59

RECORD LAYOUT AND TEST DATA

RECORD LAYOUT AND TEST DATA

61

RECORD LAYOUT AND TEST DATA

RECORD NO.

RECORD DESCRIPTOR

DESIGNING REPORTS

OBJECTIVE

To design the print charts for two reports printed on **preprinted forms** and one report printed on stock paper.

KEY WORDS

If you are unfamiliar with any of the following terms or phrases, it is suggested that you look them up in the glossary of your text or in a data processing dictionary.

Form 941a   Print charts
Preprinted forms   Templet

PREPARATION

Review the material in Chapter 6 (Detailed Design: Output) of *Systems Analysis and Design* (or its equivalent) on the design and layout of reports.

Materials needed:

1. TEMPLET. Make sure your templet has the inches divided into tenths.

2. W-2 and FICA report forms (provided).

3. Print charts (provided).

General comments regarding the completion of the report layouts:

1. The continuation sheet for Schedule A of **form 941a** is reprinted on page 66. The first page of the report shows the total number of pages and the total amount that must be included with the report to cover the employees' contribution for FICA. You need not be concerned with page 1.

2. In laying out the FICA report, use your templet to determine what print positions will be used to supply the variable information. Assume that your printer has the standard 10 print positions per horizontal inch. Vertically 8 lines per inch are printed. Use X's to indicate where the variable data will be printed. You need not add the constant data preprinted on the form.

3. You will also need your templet to determine what print positions will be used to supply the variable information for the W-2 form.

4. For the payroll register report, use the following standards.

    a. Appropriate headings must be printed on each page. The first heading, which supplies the name of the report, must have the page number and date.

    b. Column headings must be printed to indicate what variable data is printed on the report.

    c. On the print charts, the constants will be printed and the location of the variable data will be indicated by using X's.

    d. Print charts are retained as part of the documentation.

5. Study the stock deduction report on the following page and observe the conventions used.

TASKS

1. Do the print chart for the FICA report.
2. Do the print chart for the W-2 form.
3. Do the print chart for the payroll register report.

Fold back at dot

```
                              Stock Deduction Report      MM/DD/YYYY    Page NN

                                            Units    Total  Unit       Total
              Employee            Acct.Bal Purchased Units  Value      Value
     XXXXXXXXXXXXXXXXXXXXXXXXXX     XXX.XX    XXX    XXXXX  XXX.XX  X,XXX,XXX.XX




     Totals                      XXXXX,XXX.XX   XXXXXX  XXXXXXXX   XXX.XX XXXX,XXX.XX
```

Fold back at d

FORM 941a (Rev. July 1971) "A"
Department of The Treasury
Internal Revenue Service

CONTINUATION SHEET FOR SCHEDULE A OF FORMS 941, 941—M, 941SS, OR 943
REPORT OF WAGES TAXABLE UNDER THE FEDERAL INSURANCE CONTRIBUTIONS ACT

Date Quarter Ended	Page Number

If this form is used as a continuation sheet for Form 943, Employer's Annual Tax Return for Agricultural Employees, please check here. ▶

See Instructions in Publication No. 393.
Attach only original continuation sheets to your tax return. Do not send a carbon copy to the Internal Revenue Service.

STATE IDENT. NO.

Type or print in this space employer's identification number, name, and address exactly as shown on the return.

FEDERAL IDENT. NO.

EMPLOYEE'S SOCIAL SECURITY ACCOUNT NUMBER (If number is unknown, see Circular E) 000 00 0000	NAME OF EMPLOYEE (Please type or print)	TAXABLE F.I.C.A. WAGES Paid to employee in Quarter (Before deductions) ▼ Dollars Cents	TAXABLE TIPS REPORTED (See instructions Form 941) Dollars Cents

TOTALS FOR THIS PAGE
number of employees, taxable wages and tips.

Number of Employees	$	$

FEDERAL GOVERNMENT COPY

31-0455440

Form 1

1 Control number		2 Employer's State number
	22222	

3 Employer's name, address, and ZIP code

4 Subtotal ☐ Correction ☐ Void ☐

5 Employer's identification number

6 Advance EIC payment　　　　7

8 Employee's social security number	9 Federal income tax withheld	10 Wages, tips, other compensation	11 FICA tax withheld	12 Total FICA wages

13 Employee's name (first, middle, last)

14 Pension Plan Coverage? Yes/No　15　　16 FICA tips

18 State income tax withheld	19 State wages, tips, etc.	20 Name of state
21 Local income tax withheld	22 Local wages, tips, etc.	23 Name of locality

WAGE AND TAX STATEMENT

COPY B TO BE FILED WITH EMPLOYEE'S FEDERAL TAX RETURN

FORM W-2 THIS INFORMATION IS BEING FURNISHED TO THE INTERNAL REVENUE SERVICE　　DEPARTMENT OF THE TREASURY-INTERNAL REVENUE SERVICE

APP. 1979

Form 2

1 Control number		2 Employer's State number
	22222	

3 Employer's name, address, and ZIP code

4 Subtotal ☐ Correction ☐ Void ☐

5 Employer's identification number

6 Advance EIC payment　　　　7

8 Employee's social security number	9 Federal income tax withheld	10 Wages, tips, other compensation	11 FICA tax withheld	12 Total FICA wages

13 Employee's name (first, middle, last)

14 Pension Plan Coverage? Yes/No　15　　16 FICA tips

18 State income tax withheld	19 State wages, tips, etc.	20 Name of state
21 Local income tax withheld	22 Local wages, tips, etc.	23 Name of locality

WAGE AND TAX STATEMENT

COPY B TO BE FILED WITH EMPLOYEE'S FEDERAL TAX RETURN

FORM W-2 THIS INFORMATION IS BEING FURNISHED TO THE INTERNAL REVENUE SERVICE　　DEPARTMENT OF THE TREASURY-INTERNAL REVENUE SERVICE

APP. 1979

Form 3

1 Control number		2 Employer's State number
	22222	

3 Employer's name, address, and ZIP code

4 Subtotal ☐ Correction ☐ Void ☐

5 Employer's identification number

6 Advance EIC payment　　　　7

8 Employee's social security number	9 Federal income tax withheld	10 Wages, tips, other compensation	11 FICA tax withheld	12 Total FICA wages

13 Employee's name (first, middle, last)

14 Pension Plan Coverage? Yes/No　15　　16 FICA tips

18 State income tax withheld	19 State wages, tips, etc.	20 Name of state
21 Local income tax withheld	22 Local wages, tips, etc.	23 Name of locality

WAGE AND TAX STATEMENT

COPY B TO BE FILED WITH EMPLOYEE'S FEDERAL TAX RETURN

FORM W-2 THIS INFORMATION IS BEING FURNISHED TO THE INTERNAL REVENUE SERVICE　　DEPARTMENT OF THE TREASURY-INTERNAL REVENUE SERVICE

APP. 1979

SHARE PRINT CHART   PROG. ID._____   PAGE _____
(SPACING: 6 LINES PER INCH, DEPTH: 51 LINES)   DATE _____

PROGRAM TITLE _____

PROGRAMMER OR DOCUMENTALIST: _____

CHART TITLE _____

Fold back at dc

Fold back at dc

© 1981 SRA

SHARE PRINT CHART

PROG. ID. _____

PAGE _____

DATE _____

(SPACING: 6 LINES PER INCH. DEPTH: 51 LINES)

PROGRAM TITLE _____

PROGRAMMER OR DOCUMENTALIST: _____

CHART TITLE _____

DESIGNING SOURCE DOCUMENTS

To design a source document to be used in making the necessary changes to the payroll master file.

Review the material in Chapter 7 (Detailed Design: Input) of *Systems Analysis and Design* (or its equivalent) on the design of source documents.

General comments regarding the completion of the assignment:

1. The general format for entering the data into PAMØØ3 is as follows:
   a. Each field of data that might be changed during the execution of PAMØØ3 will have an identifying number.
   b. If the field is to be changed, the payroll clerk will check the number and print the data to be entered on the source document.
   c. On the source document, each field will have its own code and area for entering the new data.
   d. When the program is being executed, the dialogue between the computer and the operator will be as follows:

   (1) COMPUTER: ENTER EMPLOYEE NUMBER   / / / / /
       OPERATOR: 02360

       COMPUTER: RECORD IS FOR MARK SMITH   CONFIRM: Y/N
       OPERATOR: Y[1]

   (2) COMPUTER: ENTER NUMBER OF FIELD   / /
       OPERATOR: 01

       COMPUTER: FIELD TO BE CHANGED IS
                 EMPLOYEE'S NAME
                 ENTER FIRST NAME   / / / / / / / / / /
       OPERATOR: MARY

---

[1]If the operator responds with an N, program branches to (1).

	COMPUTER:		CONFIRM: Y/N
	OPERATOR:		N[2]
(3)	COMPUTER:	ENTER LAST NAME	/ / / / / / / / / /
	OPERATOR:		DAMUTH
	COMPUTER:		CONFIRM: Y/N[3]
	OPERATOR:		Y
(4)	COMPUTER:	ENTER MIDDLE INITIAL	/
	OPERATOR:		A
	COMPUTER:		CONFIRM: Y/N[4]
(5)	COMPUTER:	MORE CHANGES FOR MARK SMITH?	CONFIRM: Y/N[5]
	OPERATOR:		Y
(6)	COMPUTER:	MORE RECORDS TO BE CHANGED?	CONFIRM: Y/N
	OPERATOR:		Y[6]

2. The employee-number field cannot be changed.

3. The following fields within an employee's master file record can be changed by PAM002.

Name	First, last, and middle initial
Addresses	First 3 lines permit up to a 25 character address; zip code is five digits.
Employee status	S, H, C, P
Salary	Annual salary
Contract	Amount and number of payments
Hourly employees	Pay code
Bi-weekly deductions for:	Stock, credit union, union dues, term insurance, and United Fund
United Fund	Number of installments
Employee tax credit	Bi-weekly amount
Social security number	
Exemptions	
Cost center and account number	
Termination date	

TASK   Design the source document needed for PAM003.

---

[2]If the operator responds with an N, program branches to (2).
[3]If the operator responds with an N, program branches to (3).
[4]If the operator responds with an N, program branches to (4).
[5]If the operator responds with a Y, program branches to (2).
[6]If the operator responds with a Y, program branches to (1).

FORMATTING A CRT SCREEN

To format data to be displayed on a CRT.

If you are unfamiliar with the following terms or phrases, it is suggested that you look them up in the glossary of your text or in a data processing dictionary.

ASCII code
Leading zeros
Password
Prompt
VT-52 DECSCOPES

Review the information on procedures PAMØØ1, PAMØØ2, and PAMØØ3. You will also need to review the material you developed for the input requirements for each program and the material provided in assignment 6.

General comments regarding the completion of the assignment:

1. In the design of the three programs, Walczak and Arnold have used the following conventions:

   a. At the beginning of each program, terminal operators are asked if they want instructions. If they reply with a Y, the following instructions are displayed on the screen.

      (1) THE NAME OF EACH FIELD OF DATA TO BE ENTERED IS DISPLAYED.

      (2) KEY IN THE DATA RECORDED ON THE SOURCE DOCUMENT FOR THE FIELD.

      (3) PRESS THE RETURN KEY.

     (4) CONFIRM: Y/N WILL THEN APPEAR ON THE CRT.

     (5) VISUALLY CONFIRM THE DATA YOU KEYED IN.

     (6) IF CORRECT: KEY IN A Y. PRESS THE RETURN KEY. THE NAME OF THE NEW FIELD OF DATA TO BE ENTERED IS DISPLAYED.

     (7) IF INCORRECT: KEY IN A N. PRESS THE RETURN KEY. THE NAME OF THE FIELD OF DATA YOU ENTERED INCORRECTLY WILL BE DISPLAYED ON THE CRT.

     (8) AFTER ALL FIELDS OF DATA HAVE BEEN ENTERED, THE FOLLOWING MESSAGE IS DISPLAYED:

        IF MORE DATA IS TO BE ENTERED, REPLY Y.

     (9) ANY OTHER RESPONSE TO THE MESSAGE DISPLAY IN 8 WILL TERMINATE THE JOB.

  b. A confirmation report is printed and verified by a different payroll clerk than the one who entered the data.

  c. The **prompt** will display not only the name but the size of the field. The field size will be indicated by using slashes. A four-position field would be displayed as ////. The data is entered directly below the slashes.

  d. Because of the manner in which the program is written, **leading zeros** need not be entered. Commas are not entered; decimals must be entered.

  e. The complete documentation contains all the information the payroll clerk needs to sign on the system and to run the program. The **password** is changed periodically and the clerk must obtain the current password from Chang.

2. **VT-52 DECSCOPES** are used by Delta Products. The CRT has 80 horizontal display positions. Twenty-four lines can be displayed on the screen at one time. After the 24th line is displayed on the screen (including blank lines), the screen is cleared and the 25th line will be displayed at the top of the screen. Blank lines may be used to make the information more readable.

3. Do not format the general instructions printed for the operator.

4. In formatting PAM002 you are to assume that one line at a time will be displayed much like the dialogue that was illustrated in Section 6. However, format all the messages. Determine what messages you will want, how many characters in each message, and the length of each field before you begin to format your screen.

     Although controls are built into the system, it is possible that an employee can be paid for an incorrect number of hours. When this occurs, the employee will receive a check for the difference. An adjustment must be made to the employee's master file record. A check might also have been issued for someone who did not work during a particular pay period. In this case, the check is cancelled and the appropriate adjustments are made to the master file.

     In order to establish a clear audit trail, a report of changes will be printed that shows the incorrect amount and the correct amount. In the case of a check being issued for an employee who did not work, the incorrect amounts would be the gross pay received and all the deductions taken from the employee's gross. The correct amounts would be zeros.

     The fields that might need to be changed are:

Year-to-date:	FICA, state tax, city tax, federal tax, tax credit, stock deduction, credit union, term insurance, United Fund, union dues, and gross earnings.
Quarterly:	Gross, FICA, federal tax, city tax, and state tax.
Year-to-date earnings:	Either the amount in the salaried, hourly, part-time, or contract field.
Miscellaneous:	Stock deduction and stock balance fields, United Fund times paid, compensatory hours, and date hired.

The source documents have the codes to be used in making the required changes and space for recording both the wrong and right amounts.

5. Make certain the prompt contains the right name for the field of data and the correct number of slashes are displayed. The code entered could be used to retrieve the field name and correct number of slashes from a table. This would permit a loop to be used when adjusting quarterly, year-to-date, and the miscellaneous field.

6. One of the primary concerns is the ease with which the operator can read the information displayed on the screen. Any of the standard ASCII characters can be displayed on the VT-52 screen. As an example, a portion of the screen format for PAMØØ1 is provided on the following page.

7. In formatting the screen for PAMØØ3, assume that Delta Products uses a technique which permits the entire format to be displayed at one time. Variable data is displayed as the operator enters the employee's name and change code. If an amount is entered incorrectly, the operator can control the cursor and make the required correction in the amount displayed on the CRT.

   The changes made using PAMØØ3 are the ones that normally occur and are not the result of an error. The fields that can be changed are: name, addresses 1, 2, and 3, zip code, employee's status, date terminated, salary, contract, contract pay periods, stock deduction, credit union, term insurance, United Fund amount and times, union dues, tax credit, social security number, tax status, number of exemptions, city code, cost accounting numbers, and pay code.

TASKS

1. Answer the questions on page 79.
2. Using the forms provided, format the screen for PAMØØ2.
3. Using the forms provided, format the screen for PAMØØ3.

DELTA PRODUCTS/CRT DISPLAY LAYOUT

Job Number: **PAM ØØ1**   Programmer: **WALCZAK**   Date: **06/10/81**

POSITION

Line	Content
03	NAME - FIRST ///////////
04	
05	
06	
07	CONFIRM: Y/N
08	
09	LAST /////////////////
10	
13	CONFIRM: Y/N
15	MIDDLE INITIAL
16	/
19	CONFIRM: Y/N
21	ADDRESS:
22	LINE 1
23	///////////////////////////////
26	CONFIRM: Y/N
28	LINE 2
29	///////////////////////////

NOTE: (NOT PART OF FORMAT)

OPERATOR ENTERS DATA BELOW

THE SLASHES. NAME - FIRST

WILL DISPLAY AND THE SLASHES.

AFTER RETURN KEY IS DEPRESSED

CONFIRM: Y/N IS DISPLAYED

UP TO 4 LINES MAY BE USED.

ZIP CODE MUST BE ON FOURTH LINE.

USE RETURN KEY FOR A BLANK LINE.

LINE

THOUGHT STARTERS

1. What correlation regarding the field names should exist between the source document, master file layout, display on the CRT, and source code for the computer program?

   _____

   _____

   _____

   _____

   _____

   _____

   _____

   _____

2. According to Delta Products' programming standards, what factors should be considered when determining the name for each field?

   _____

   _____

   _____

   _____

3. If leading zeros need not be entered, why are the slashes used to designate the size of the field?

   _____

   _____

4. In the documentation, the operator is instructed not to use commas. However blanks are to be inserted in place of the commas. To do this the operator pressed the space bar. Decimals are to be entered. Why did Walczak and Arnold instruct the operator to enter blanks in place of commas? The program will delete blanks in a numeric field; blanks in alphanumerical fields will be retained.

   _____

   _____

   _____

   _____

DELTA PRODUCTS/CRT DISPLAY LAYOUT

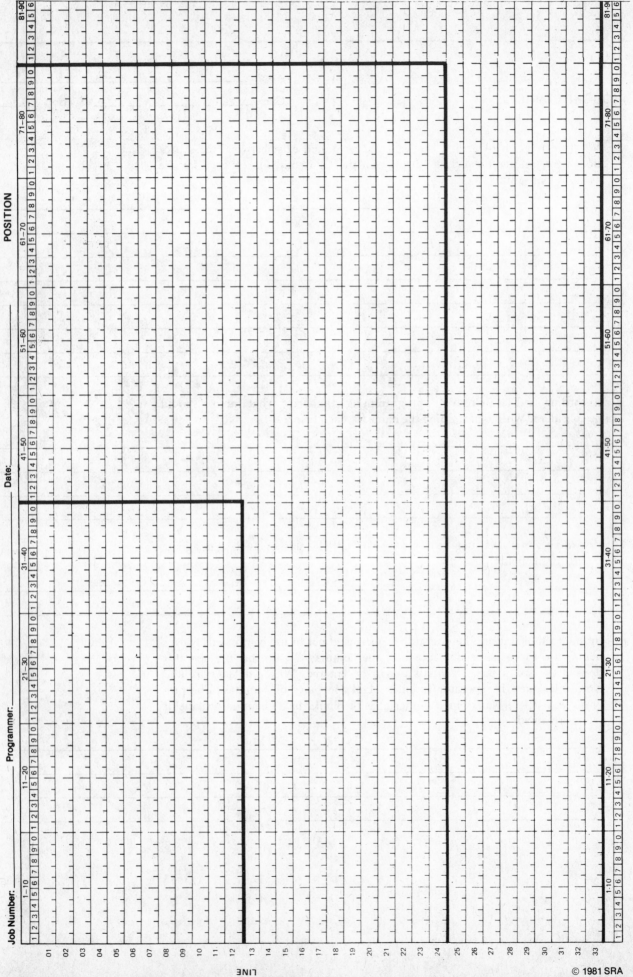

DELTA PRODUCTS/CRT DISPLAY LAYOUT

Job Number: _____ Programmer: _____ Date: _____

POSITION

DELTA PRODUCTS/CRT DISPLAY LAYOUT

POSITION

Job Number: _____ Programmer: _____ Date: _____

LINE

DELTA PRODUCTS/CRT DISPLAY LAYOUT

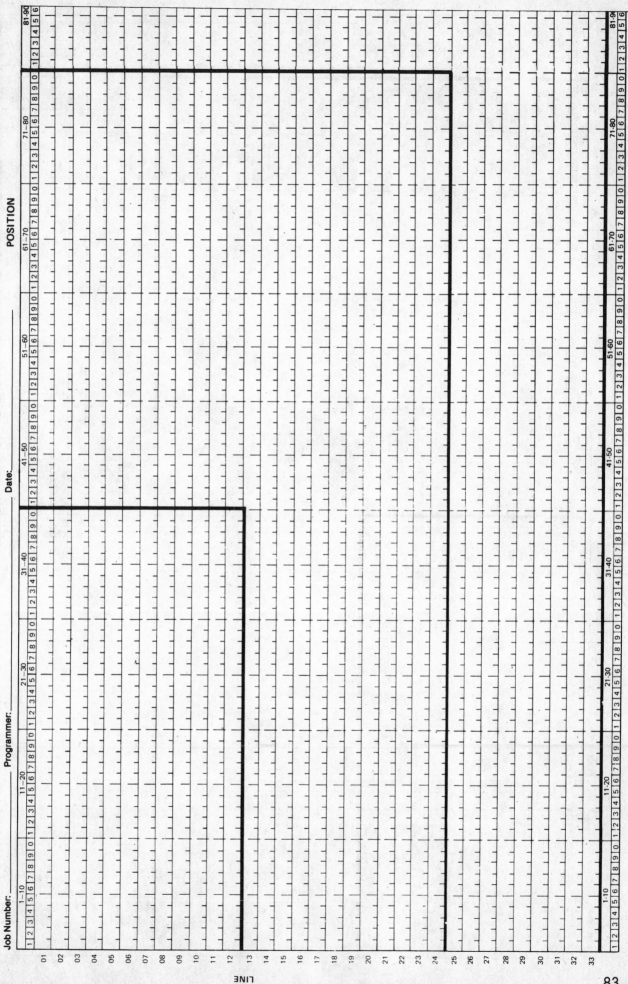

83

DELTA PRODUCTS/CRT DISPLAY LAYOUT

Job Number: _____

Programmer: _____

Date: _____

POSITION

LINE

© 1981 SRA

8

INPUT, PROCESSING, AND OUTPUT CONTROLS

To determine the input, internal program, and output controls for each procedure. **OBJECTIVE**

Review the material in Chapters 6–10 of *Systems Analysis and Design* (or its equivalent) on input, internal program, and output controls. **PREPARATION**

General comments regarding the completion of the assignment:

1. The chart on page 86 provides one example of how the assignment can be completed. Your instructor may prefer that you use a different format.

2. As you do the assignment you must refer to Section 4 to determine what input is being entered and what output will be printed or written into a file.

3. All data entering the master file must be edited and/or confirmed either as it enters the system or by some type of printed report that is audited by the payroll department. Therefore, once the master file is created you may assume that it contains valid data. However, some of the data might be used in editing new data as it enters the system.

4. All fields of data transferred to the current files were edited (if possible) as they entered the system, transferred from the master file, or computed. Therefore, when a current file is used as input to a report-producing program, the fields of data need not be edited.

Using a format similar to the one illustrated on the next page, indicate what controls **TASKS** would be used for each of the payroll programs listed on pages 50–51.

CONTROLS

Program Number	Input	Processing	Output
PAMØØ1	Visually confirm all data entered from the terminal Numeric fields edited to make certain only digits entered	Account number—greater than 10; less than 20 Cost center—greater than 0; less than 51 Employee status—S, H, C, or P Salaried employees have salary entered Hourly workers have pay code Contract employees have amount of contract and number of installments. Social security number—no blanks, 9 digits entered United Fund—greater than 0; installments entered Tax status—M or S City code—1 if within city; 0 if out of the city Zip code—numeric data Add 1 to records created	Confirm total of records entered with batch total Visually verify printed report for factors such as correct spelling of names, accuracy of numeric data
PARØØ9	All data in current earnings file has been validated	Accumulated totals	Confirm total checks printed with record count for PARØØ4 Check totals printed at EOJ with the totals printed on the payroll register

9

DEVELOPING A HIERARCHY CHART AND A STRUCTURED PROGRAM FLOWCHART

To develop a **hierarchy chart** and a **structured program flowchart** for PARØØ4.

If you are unfamiliar with the following terms or phrases, it is suggested that you look them up in the glossary of your text or in a data processing dictionary.

ANSI flowchart symbols	Modules
Function	Nassi-Schneiderman charts
Hierarchy chart	Processing module
HIPO chart	Pseudocode
Indicator	Structured program flowchart
Initialization module	Submodules
IPO chart	Termination module
Line counter	VTOC
Main control module	

1. Read the material that follows.

2. Read Chapter 10 (Programming Considerations) of *Systems Analysis and Design* (or its equivalent).

3. Review the material on the payroll system to make certain that you understand how employees are paid.

4. Review the use of flowcharting symbols.

General comments regarding the completion of the assignment:

1. Delta Products uses a structured approach to programming and follows the guidelines presented below.

87

    a. Each module has one function. However in a print module the following tasks would be performed: data is moved to the print structure; the **line counter** is checked to see if headings are to be printed; and the detail line is printed. In a write module, data is moved to the output structure, and the record written into the desired medium.

    b. For each file there will be only one read or write statement. However more than one print statement will be required (depends on language being used).

2. The payroll master file and the transaction file that contains the time card images are accessed sequentially.

3. A separate module should be used to compute the gross pay for hourly or part-time, salaried, and contract employees.

4. The employees who have transaction files are:

    a. hourly employees;

    b. part-time employees;

    c. salaried workers with compensatory time; and

    d. salaried workers who are not to be paid.

5. The employees might have more than one transaction record.

6. All the data was edited by PARØØ3 or when the data was written in the payroll master file. Therefore the input need not be edited.

7. A subroutine stored in an online library is used for computing the federal, state, and city tax.

8. When moving data from an input structure to an output structure, do not detail all the fields to be moved. Show one block on the flowchart which has the caption "move data to current earnings file."

9. Use the following five **indicators** in your flowchart.

    a. The READ TRAN indicator is initially set to No. When a match is found (employee number from the transaction file is equal to the employee number from the payroll master file) the READ TRAN indicator is set to Yes. After a transaction record is read, READ TRAN is set to No.

    b. The PROCESS RECORD indicator is initially set to Yes. When there is a salaried employee with a no-pay record or an hourly or part-time employee without a transaction record, the PROCESS RECORD is set to No. Each time a new payroll master file record is read, the PROCESS RECORD indicator is set to Yes.

    c. The READ MASTER indicator is initially set to No. If the employee number in the transaction file is not the same as the employee number in the master file, the READ MASTER indicator is set to Yes. If the two numbers are the same, READ MASTER is set to No.

    d. The MORE RECORDS indicator is initially set to Yes. When the end of the transaction file is detected, the MORE RECORDS indicator will be set to No.

    e. The WRITE PARTIAL indicator is initially set to No. When a transaction record with an M code is read, the WRITE PARTIAL indicator is set to Yes. When the indicator has a value of Yes, a partial current earnings record will be written and much of the normal processing will be by-passed. When a new transaction record is read, WRITE PARTIAL will be reset to No.

10. In the initialization module, the modules that read the transaction file and the payroll master file will be envoked. The transaction file must be read prior to reading the payroll master file.

11. The job will terminate when the end of the payroll master file is detected. When the end of the transaction file is detected, the READ TRAN is set to No and normal processing continues.

12. The program will be written in COBOL. Therefore PERFORM . . . UNTIL is used to set up the main process-payroll-records loop.

TASKS

1. Answer the questions on page 95 regarding the flowchart for PARØØ4.
2. Prepare a hierarchy chart for PARØØ4.
3. Prepare a modular flowchart for PARØØ4.

DEVELOPING A HIERARCHY CHART

In developing the logic plan for a program, a series of steps must be completed. Before the hierarchy chart was prepared for the stock deduction program, the following documents were studied:

1. the report layout form for the stock deduction report; and
2. the master file layout.

After studying the documents, the programmer or analyst must determine what additional information is needed to produce the required results. Can the information be calculated or is a date control card or transaction file needed? In the illustration, all the data except the date and the current value of the stock is either available in the master file records or can be calculated.

Next, the analyst must think about what type of editing, error routines, and controls must be built into the program. A good analyst or programmer provides error routines for those it-can-never-happen errors. Once the analyst determines what must be done, the required tasks are grouped in **modules**.

Although there may be more, the following three major **submodules** are usually under the **main control module**.

1. *Initialization.* What must be done at the beginning of the program before a record can be processed? Included in the **initialization module** are tasks such as opening the files, loading data into a table, entering the date, and reading the first master or transaction record.

2. *Processing.* The **processing module** usually controls the logic of the program. Depending upon the complexity of the program, many submodules will be controlled by the processing module.

3. *Termination.* What tasks must be completed after all transaction records have been processed? Usually the **termination module** includes tasks such as printing totals, printing an error report, and closing files.

Although many analysts feel that each **function** should be contained in its own module, exceptions are sometimes made. In determining what modules are needed, the guidelines established for the company should be followed. Generally a module should be limited to fifty lines of code.

After the analyst has listed the tasks to be completed and grouped the tasks into functional modules, the hierarchy chart can be prepared. Study the hierarchy chart for the stock deduction program and you will observe:

Hierarchy chart for the stock deduction program

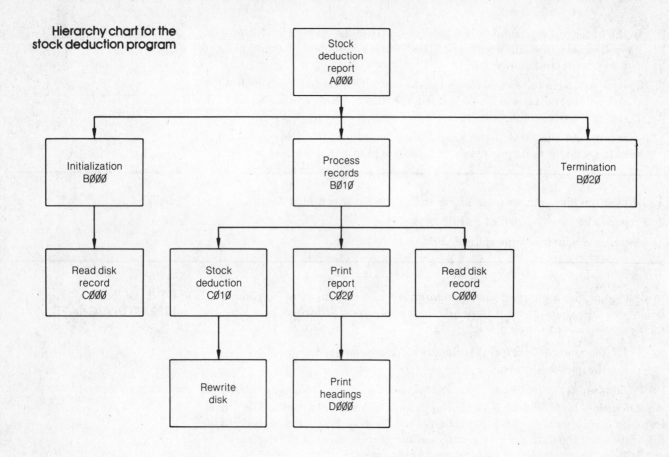

1. Each module is given a name. Depending upon the language being used, the module name on the hierarchy chart will either be used as a comment or paragraph name in the program.
2. The AØØØ module envokes the BØØØ, BØ1Ø, and BØ2Ø modules.
3. The BØØØ module envokes the CØØØ module; C modules envoke D modules.
4. Since the Read Disk Record can be envoked by both the BØØØ and BØ1Ø modules, the top right corner of the symbols is shaded.

The hierarchy chart is a plan, or visual table of contents (**VTOC**), which should be followed in developing a program flowchart. Although some programmers prefer to use **pseudocode, HIPO charts, IPO charts,** or **Nassi-Schneiderman charts** for developing the detailed logic, the same thought process must precede the development of the logic plan. The program must be divided into meaningful modules and submodules.

DEVELOPING A PROGRAM FLOWCHART

The programmer developing the plan should use the standard **ANSI flowchart symbols**. Often the question is asked regarding the amount of detail that should be included in the program flowchart (or other logic plan). While experienced programmers tend to leave out some of the routine detail, this practice is unwise since an inexperienced programmer may need to follow existing flowcharts to modify programs.

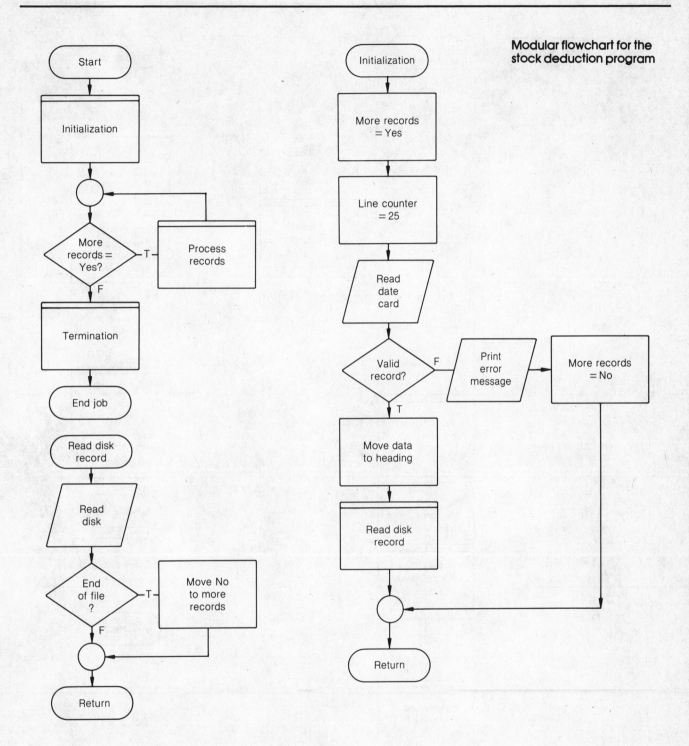

Modular flowchart for the stock deduction program

In studying the stock deduction flowchart, you will observe that each module listed on the hierarchy chart has its own flowchart. The module names used on the flowchart match the ones used on the hierarchy chart. While the main control module (stock deduction report) has start and end in the first and last symbols, the submodule's first and last symbols contain the name of the module and return. When a submodule is envoked, control should always return to the envoking module.

Modular flowchart for the stock deduction program (continued)

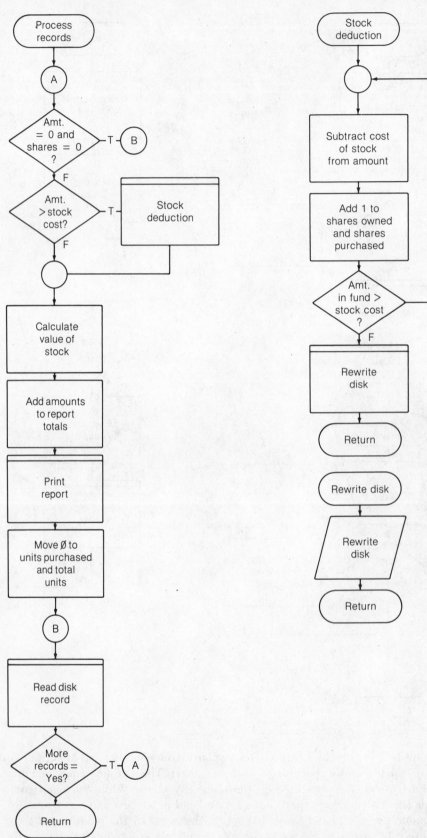

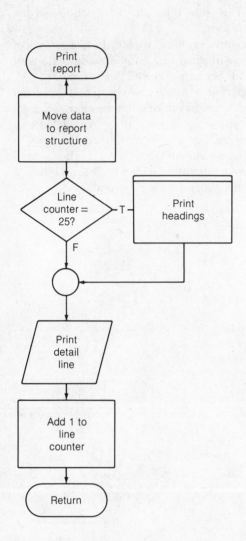

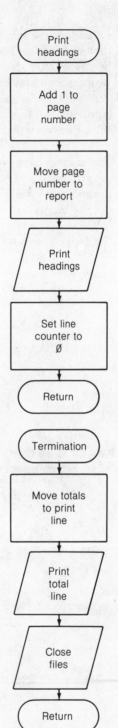

CONCLUSIONS REGARDING HIERARCHY AND PROGRAM FLOWCHARTS

Often the first hierarchy chart developed by the analyst, as well as the logic plan that supports the hierarchy chart, will need to be modified. The modification may be made after the design walkthrough. The team members who conducted the walkthrough may have found errors or omissions.

After the design walkthrough, the programmer writes the source code. Although the flowchart should be language-independent, minor modifications in the logic plan will sometimes need to be made because of the language used.

Perhaps the greatest value in preparing a hierarchy chart and program flowchart is completing the thought process necessary to develop the charts. Once a programmer gets into the habit of listing the tasks or steps to be completed and grouping the tasks into meaningful modules, he or she is well on the way to developing well-structured programs. A well-developed logic plan can also be used in the design walkthrough. If omissions and errors in logic can be detected during the design phase, far less time will be spent in debugging the program.

THOUGHT STARTERS

1. Why was it necessary to use a READ MASTER indicator?

   _____

   _____

   _____

2. At the end of the main process records loop, the analyst indicated that in all cases unconditionally both a transaction and a payroll master file record were to be read. What would occur if this error were not detected?

   _____

   _____

   _____

   _____

   _____

3. Why is it necessary to use a PROCESS RECORD indicator?

   _____

   _____

   _____

   _____

4. What will occur if the PROCESS RECORD indicator is equal to No?

   _____

   _____

5. When will the READ TRAN indicator be set to No?

   _____

   _____

   _____

   _____

6. What is the purpose of a hierarchy chart?

   _____

   _____

7. Why might an analyst revise the hierarchy chart after the more precise development of the program logic is begun?

    _____

    _____

    _____

10

DOCUMENTING THE SYSTEM

To document portions of the payroll system.   **OBJECTIVE**

If you are unfamiliar with the following terms or phrases, look them up in the   **KEY WORDS**
glossary of your text or in a data processing dictionary.

Run manual cover sheet
Run sheet
Systems flowchart

1. Read Chapter 12 (Implementation: Documentation and Evaluation) of *Systems*   **PREPARATION**
   Analysis and Design (or its equivalent).
2. Reread sections 4–9 to review the general concepts of the payroll system.

General comments regarding the completion of the documentation assignment:

1. Make certain that you understand each of the payroll procedures described on
   pages 50 and 51.
2. Keep in mind for whom the documentation is intended.
3. Remember that a good test for documentation is to have someone unfamiliar
   with the system or task read the material you have prepared.
4. In completing this assignment, assume that you are an analyst employed by
   Delta Products. In evaluating your performance, Ben Paul will consider the
   quality of your documentation.
5. Many of the items included in the documentation for PARØØ4 have already been
   completed. You may wish to copy some of the previous assignments to make
   necessary corrections and to prepare higher quality documentation.

TASKS

1. Write an overview of the payroll system. The overview will be used by the payroll department employees, management, other members of the data processing department, and auditors to obtain an understanding of how the system functions. Excessive detail should not be included.

2. Document PAMØØ4 according to the following directions:

Run manual cover sheet	Use the form provided.
Overview	Include a **systems flowchart** at the top of the page followed by a written overview of the procedure.
I/O formats	Prepare record layouts for the payroll master, transaction, current earnings, stock deduction, credit union, weekly earnings, and income tax credit files.
Controls	Provide a written overview of the controls designed into PAMØØ4.
Hierarchy chart	Make any corrections necessary to the one submitted in Section 9.
Program flowchart	Make any corrections necessary to the one submitted in Section 9.
Run sheet	Use the form provided and complete the run sheet.

In preparing the run sheet include specific directions on when the program is to be run. Keep in mind an audit run is performed prior to the final run on two-part paper. Assume that a removable disk pack contains the payroll files. All the files are on the same pack. The job control language required to execute the job is stored in a library and is referenced by SLPARØØ4.

d delta products RUN MANUAL COVER SHEET

PROGRAM NAME			LANGUAGE USED
JOB NUMBER	PROGRAMMER		DATE

EXECUTION PROBLEMS: CONTACT:

REVISIONS:

By	Date	Authorized by	Reason

RECORD LAYOUT AND TEST DATA

RECORD NO.

RECORD DESCRIPTOR

© 1981 SRA

RECORD LAYOUT AND TEST DATA

101

RECORD LAYOUT AND TEST DATA

RECORD NO.

RECORD DESCRIPTOR

d delta products PRODUCTION RUN INSTRUCTIONS

PROGRAM NAME		FREQUENCY

TYPE OF RUN
○ MAINTENANCE   ○ REPORT

JOB NUMBER	PROGRAMMER	DATE	PUNCH

SOURCE ○   OBJECT ○   CATALOG ○

DATE CARD

DISTRIBUTION

FORM TYPE

SPACING
○ 6LPI   ○ 8LPI

CARRIAGE: ○ STANDARD
○ OTHER _____

DISK OR TAPE FILE NAME	ADDRESS	SPECIAL INSTRUCTIONS

SPECIAL INSTRUCTIONS:

PROGRAMMED MESSAGES

CONTROLS

d delta products PRODUCTION RUN INSTRUCTIONS

PROGRAM NAME		FREQUENCY

		TYPE OF RUN
JOB NUMBER	PROGRAMMER	DATE
		○ MAINTENANCE ○ REPORT
		PUNCH

DATE CARD	SOURCE ○ OBJECT ○ CATALOG ○
	DISTRIBUTION
	FORM TYPE
	SPACING ○ 6LPI ○ 8LPI
	CARRIAGE: ○ STANDARD ○ OTHER _____

DISK OR TAPE FILE NAME	ADDRESS	SPECIAL INSTRUCTIONS

SPECIAL INSTRUCTIONS:

PROGRAMMED MESSAGES

CONTROLS

11

IMPLEMENTING THE SYSTEM

1. To determine the optimum time to implement the new payroll system.
2. To determine the tasks needed to implement the new system.

OBJECTIVES

1. Review the materials pertaining to the payroll system. Consider the reports needed and the content of the various files.
2. Review Chapter 11 (Implementation: Preparing for the New System) of *Systems Analysis and Design* (or its equivalent).

PREPARATION

General comments regarding the assignment:

1. In completing the assignment, consider the tasks performed by the payroll clerks, accountants, data processing personnel, and cost-center supervisors under the old and new systems.
2. When the system becomes operational, documentation and training of employees must be completed.
3. Delta Products does not have a training department. The graphic arts department can assist in the preparation of some audio-visual materials.

1. Answer the questions on pages 107 and 108.
2. List all the tasks that must be performed to implement the new system.
3. Sequence the tasks in the order that they must be performed. If two or more tasks can be completed simultaneously, give the tasks the same number.

TASKS

THOUGHT STARTERS

1. What would be the optimum time to start using the new system to process the payroll data?

   _____

   Why? _____

   _____

   _____

2. What type of conversion would you recommend?

   _____

   Why? _____

   _____

   Explain exactly how you would convert to the new system?

   _____

   _____

   _____

   _____

3. Could the payroll master file be created in October if the system is to become operational for the first pay in January? Explain your answer.

   _____

   _____

4. Would you recommend that a special program be written to take the data stored in the old payroll master file and convert it to the format required for the new payroll system?

   _____

   Why do you feel converting the master file in this manner would be a good, or poor idea?

   _____

   _____

   _____

5. If you did write a special program to be used to convert the data stored in the old master file to the new format, how would you enter the fields of data that were not required under the new system?

_____

_____

_____

6. What reports will be printed on stock paper?

_____

_____

7. What reports will be printed on preprinted forms?

_____

_____

8. Why would it be necessary to redesign the payroll register report?

_____

_____

12

PROVIDING IN-SERVICE TRAINING

OBJECTIVES

1. To determine which employees' jobs will be affected by the new payroll system.
2. To determine what type of training program needs to be developed to prepare employees to work with the new system.

PREPARATION

1. Review the portions of Chapter 11 (Implementation: Preparing for the New System) that pertain to training employees.
2. Review the material on the old and new payroll systems.

General comments regarding the assignment:

1. Throughout the investigation and design phases of the project, management and the users were directly involved.
2. Whenever new reports or source documents were designed, users were asked to provide input and to approve the final design.
3. The management of Delta Products believes that before any major system is made operational, general education pertaining to the new system should be provided.
4. Delta Products has a graphic arts department that assists in the preparation of overhead transparencies, slides, slide/tape presentations, and charts.
5. A monthly Delta Products news bulletin is printed and distributed to all employees.
6. In certain cases it might be better to work with some individuals on a one-to- one basis rather than providing group instruction.

TASKS

1. Make a list of all individuals directly affected by the new system. Review the information provided regarding the old and the new payroll systems. What individuals outside of the payroll and data processing departments will be directly involved with the new system.

2. Make a list of individuals who will be indirectly affected by the new system.

3. For each individual (or group) on your list, indicate how he or she will be affected by the new system and what type of training should be provided. Also indicate if the conversion to the new system would require a major or minor adjustment.

4. Prepare your report in a manner similar to the one illustrated below.

Individual or Group	Adjustment	DIRECT INVOLVEMENT	
		Change Required	Training to be provided and by whom
Keypunch operators	Minor	1. Less data to be prepared since terminals will be used for inputting some of the data into the new system. 2. New formats for some of the cards that will be keypunched and verified.	1. The operations manager, Betty Nichols, has been directly involved in the redesign of the payroll system and is aware of the changes. 2. Review the changes with Betty Nichols. 3. Documentation prepared for the keypunch operators by Betty Nichols. 4. Keypunch operators will use the documentation to learn the new formats.